D0072223

DYSGENICS

Recent Titles in
Human Evolution, Behavior, and Intelligence

Staying Human in the Organization: Our Biological Heritage and the Workplace
J. Gary Bernhard and Kalman Glantz

The End of the Empty Organism: Neurobiology and the Sciences of Human Action
Elliott White

Genes, Brains, and Politics: Self-Selection and Social Life
Elliott White

How Humans Relate: A New Interpersonal Theory
John Birtchnell

Choosing the Right Stuff: The Psychological Selection of Astronauts and Cosmonauts
Patricia A. Santy

Hormones, Sex, and Society: The Science of Physicology
Helmuth Nyborg

A People That Shall Dwell Alone: Judaism as a Group Evolutionary Strategy
Kevin MacDonald

DYSGENICS

Genetic Deterioration in Modern Populations

RICHARD LYNN

Human Evolution, Behavior, and Intelligence
SEYMOUR ITZKOFF, Series Editor

Westport, Connecticut
London

Library of Congress Cataloging-in-Publication Data

Lynn, Richard.
Dysgenics : genetic deterioration in modern populations / Richard Lynn.
p. cm.—(Human evolution, behavior, and intelligence, ISSN 1063–2158)
Includes bibliographical references and index.
ISBN 0–275–94917–6 (alk. paper)
1. Human population genetics. 2. Human genetics. I. Title.
II. Series.
GN289.L96 1996
573.2'15—dc20 96–2802

British Library Cataloguing in Publication Data is available.

Library of Congress Catalog Card Number: 96–2802
ISBN: 0–275–94917–6
ISBN: 1063–2158

First published in 1996

Praeger Publishers, 88 Post Road West, Westport, CT 06881
An imprint of Greenwood Publishing Group, Inc.

Printed in the United States of America

The paper used in this book complies with the Permanent Paper Standard issued by the National Information Standards Organization (Z39.48–1984).

10 9 8 7 6 5 4 3 2

Contents

Acknowledgments

It is a pleasure to record my appreciation for the help I have received from a number of people in writing this book. My research student, Lucy Greene, collected the data on the intelligence of children and their number of siblings described in Chapter 5, and David Farrington of the Institute of Criminology at Cambridge University supplied the data on the fertility of criminals described in Chapter 14. Edward Miller and Seymour Itzkoff read through the first draft and made a number of valuable comments. James Sabin of Greenwood Press was unfailingly helpful. I am especially indebted to Harry Weyher, president of the Pioneer Fund, for his encouragement and financial support. Betty Hemphill and Frances McLernon have provided ever-patient typing and clerical assistance. My wife, Susan, has been my severest critic, as wives should be, and has checked my logic, calculations and conclusions. To all of these I am grateful.

DYSGENICS

Chapter 1

Historical Understanding of the Problem

1. Benedict Morel Identifies Genetic Deterioration. 2. Francis Galton Formulates the Concept of Eugenics. 3. Gloomy Views of Charles Darwin. 4. Herbert Spencer and Social Darwinism. 5. Karl Pearson Restates the Problem. 6. The Rise of Eugenics. 7. Ronald Fisher on the Decay of Civilizations. 8. Julian Huxley and the Evolutionary Perspective. 9. Hermann Muller's Geneticists' Manifesto. 10. Raymond Cattell and the Decline of Intelligence. 11. The 1963 Ciba Conference. 12. The Decline and Fall of Eugenics. 13. Conclusions.

In the middle decades of the nineteenth century a number of biological and social scientists believed that the genetic quality of the populations of the Western nations was deteriorating. They thought that this was occurring because of the relaxation of natural selection, the process by which nature in each generation eliminates the unfit by reducing their fertility and by early death. Once natural selection becomes relaxed, or even ceases to operate at all, they believed, genetic deterioration would inevitably take place. They thought that this process had already begun.

Once this conclusion had been reached, some of these Victorians began to think about what could be done to counteract genetic deterioration. The person who gave most thought to this was a young cousin of Charles Darwin named Francis Galton. His solution was that natural selection should be replaced by consciously designed selection, through which human societies would control and improve their own genetic quality. For this consciously designed selection Galton (1883) proposed the term *eugenics.* Almost a century later William Shockley (1974) proposed the term *dysgenics* for the genetic deterioration that eugenics was designed to correct.

The view that the populations of Western nations were deteriorating genetically and that steps needed to be taken to correct this came to be widely accepted in the first half of the twentieth century. In the second half of the century a reaction against eugenics set in, and from the 1970s onward eugenics became virtually universally dismissed. My objective in this book is to make the case that in the repudiation of eugenics an important truth has been lost, and to rehabilitate the argument that genetic deterioration is occurring in Western populations and in most of the developing world. This opening chapter sets out the development of this idea and its subsequent rejection.

1. BENEDICT MOREL IDENTIFIES GENETIC DETERIORATION

The first full analysis of the adverse effect of the slackening of natural selection on the genetic quality of the population was made in the 1850s by a French physician named Benedict Augustin Morel (1857). Morel perceived that infant and child mortality were declining in mid-nineteenth-century France, largely as a result of improvements in public health, and consequently that many infants and children who previously would have died were now surviving to adulthood.

He argued that the increased survival rate and reproduction of the less fit must entail a deterioration of population quality. He identified the characteristics for which this was taking place as "physique" (health), "intellectuelle" (intelligence) and "morale" (moral character). Morel believed that these characteristics were transmitted in families from parents to children, through both genetic and environmental processes. He believed also that there was a degenerate class of criminals, prostitutes and paupers, a segment of society that was later to become known as the underclass, and that these groups had higher fertility than the rest of the population. He saw this as part of the problem of genetic deterioration.

Morel is little remembered today. There is no mention of him in the histories of the eugenics movement by Haller (1963), Ludmerer (1972), Kevles (1985) and Degler (1991), or in the texts on eugenics by Osborn (1940), Bajema (1976) and Cattell (1972, 1987). But, working on the basis of what was inevitably the limited evidence of the 1850s, Morel was the first to set out the essential components of the case that genetic deterioration is taking place.

2. FRANCIS GALTON FORMULATES THE CONCEPT OF EUGENICS

Morel's work does not seem to have been read by his contemporaries in England, but someone who was thinking along similar lines was Francis Galton. Galton read Darwin's *The Origin of Species* when it appeared in 1859, and he realized that the process of natural selection, by which the genetic quality of the population is maintained, had begun to weaken in the economically developed nations of the time. He first aired this problem in 1865, when he wrote that "One of the effects of civilisation is to diminish the rigour of the application of the law of natural selection. It preserves weakly lives that would have perished in barbarous lands.'' He went on to note that natural selection had weakened not only against poor health but also against low intelligence and what he called ''character'' (Galton, 1865, p. 325), the same three qualities that Morel had independently identified in France. By ''character'' Galton meant a strong moral sense, self-discipline and good work motivation. In contemporary psychology this broad trait has become known as *conscientiousness* and we shall be examining it in detail in Chapters 12 through 14. Until then I shall stick with Galton's term *character.*

Galton returned to the theme of genetic deterioration at greater length in his book *Hereditary Genius* (1869). Here he suggested that in the early stages of civilization what he called ''the more able and enterprising men'' were the most likely to have children, but in older civilizations, like that of Britain, various factors operated to reduce the number of children of these and to increase the number of children of the less able and enterprising. He thought that the most important of these factors was that able and enterprising young men tended not to marry, or only to marry late in life, because marriage and children would impede their careers. The effect of this was that ''there is a steady check in an old civilisation upon the fertility of the abler classes: the improvident and unambitious are those who chiefly keep up the breed. So the race gradually deteriorates, becoming in each successive generation less fit for a high civilisation'' (Galton, 1869, p. 414).

Galton thought that the genetic deterioration of Western populations was a serious problem and that steps needed to be taken to counteract it. In principle, he thought that this was straightforward. It would consist of replacing natural selection by consciously designed selection. This would be carried out by adopting the methods that had been used for centuries by animal and plant breeders and consisted of breeding from the best varieties to obtain improved strains. Galton proposed that the same method would work and should be applied in practice to humans.

Galton first advanced this idea in his 1865 article, and he elaborated on it in *Hereditary Genius.* Here he wrote, ''As it is easy to obtain by careful selection a permanent breed of dogs or horses gifted with peculiar powers of running, or of doing anything else, so it would be quite practicable to produce a highly gifted race of men by judicious marriages during several consecutive generations'' (Galton, 1869, p. 45). Galton researched the pedigrees of eminent men, such as lawyers, scientists and statesmen, and showed that outstanding ability and talent were frequently transmitted from generation to generation in elite families. He proposed that this was due to the genetic transmission of high ability and the character qualities of work commitment, energy and ambition. He argued that this showed it would be possible to improve the genetic quality of human populations by increasing the fertility of the talented individuals.

Galton developed this idea further in his next book, *English Men of Science.* In this he traced in greater detail the family pedigrees of a number of eminent

English scientists. He found that most of them came from the professional and middle classes and concluded that these are "by far the most productive of natural ability," although he recognized that by the process of social mobility they are "continually recruited from below," particularly from the families of skilled artisans. By contrast, he described the lower classes as "the residuum," largely devoid of the qualities of intelligence and energy necessary for high achievement (1874, pp. 9–16).

Galton continued to work on the idea that the relaxation of natural selection was causing genetic deterioration and that this could be counteracted by consciously designed selection. In 1883 he coined the word *eugenics* for the study of this set of problems. The word was constructed from the Greek to mean "good breeding." During the next three decades, Galton restated and elaborated this theme (Galton, 1883, 1901, 1908, 1909). In his autobiographical *Memoirs*, written shortly before his death in 1911, he reiterated that natural selection had broken down and that to avoid genetic deterioration it would be necessary "to replace natural selection by other processes" (1908, p. 323). He continued to affirm his view that a major part of the problem lay in the low fertility of the professional classes because "the brains of the nation lie in the higher of our classes" and that these people were having insufficient children in accordance with the general principle that "it seems to be a tendency of high civilisation to check fertility in the upper classes" (1909, pp 11, 39).

The basic elements of Galton's ideas were that there was a social-class gradient for ability and desirable character qualities of work motivation and social commitment, such that these were strongest in the professional classes and declined with falling socioeconomic status. There was also an inverse relationship between social class and fertility such that the professional class had the lowest fertility. He believed that this was leading to genetic deterioration, but that this could be corrected by measures designed to reverse the negative relationship between social class and fertility. These ideas were the essential components of eugenics.

3. GLOOMY VIEWS OF CHARLES DARWIN

One of those who read Galton's *Hereditary Genius* shortly after it was published in 1869 was Charles Darwin, and two years later Darwin discussed the problem of the relaxation of natural selection in his second major book, *The Descent of Man* (1871). Here he noted that as human societies became more civilized, they showed increasing sympathy and compassion for their weaker members. He accepted that this was ethically right but pointed out that it involved a genetic cost because it entailed the survival and reproduction of those who previously would have died or not had children. Darwin summarized the problem in these words:

> We civilised men do our utmost to check the process of elimination; we build asylums for the imbecile, the maimed and the sick; we institute poor laws; and our medical men

exert their utmost skills to save the life of everyone to the last moment. Thus the weak members of civilised societies propagate their kind. No-one . . . will doubt that this must be highly injurious to the race of man. (Darwin, 1871, p. 501)

Darwin understood that the way to prevent genetic deterioration lay in curtailing the fertility of those with socially undesirable characteristics, writing that "Both sexes ought to refrain from marriage if they are in any marked degree infirm in body or in mind" (1871, p. 918). In those days if people refrained from marriage they also, for the most part, did not have children.

A few years later Darwin talked about these problems with the biologist Alfred Russell Wallace, who had formulated the theory of evolution independently of Darwin in the 1850s, and who recorded their discussion:

In one of my last conversations with Darwin he expressed himself very gloomily on the future of humanity, on the ground that in our modern civilisation natural selection had no play and the fittest did not survive. . . . It is notorious that our population is more largely renewed in each generation from the lower than from the middle and upper classes. (Wallace, 1890, p. 93)

Wallace went on to record that Darwin spoke of the large number of children of "the scum," and of the grave danger this entailed for the future of civilization (Wallace, 1890, p. 93). Darwin understood that the relaxation of natural selection was leading to genetic deterioration.

4. HERBERT SPENCER AND SOCIAL DARWINISM

Another of those who in the 1870s understood the problem of genetic deterioration was Herbert Spencer. It was Spencer who coined the phrase "the survival of the fittest," which Darwin accepted as a useful shorthand term for the processes of natural selection by which genetically sounder or "fitter" individuals tend to survive and reproduce more than the genetically unsound or less fit. Spencer agreed with Galton and Darwin that the principle of the survival of the fittest was ceasing to operate in civilized societies, and he wrote robustly of the undesirability of this trend: "To aid the bad in multiplying is, in effect, the same as maliciously providing for our descendants a multitude of enemies. Institutions which 'foster good-for-nothings' commit an unquestionable injury because they put a stop to that natural process of elimination by which society continually purifies itself" (Spencer, 1874, p. 286).

Spencer also coined the term *Social Darwinism* for the theory that the competition between individuals for survival and reproduction, which is present throughout animal and plant species, also exists in contemporary civilizations in the form of competition for social position. He held that individuals with intelligence and the character traits of a capacity for hard work and self-discipline rose in the social hierarchy and formed a genetically elite professional and upper class, while those lacking in these qualities fell into the lower classes. He arrived at the same conclusion as Galton and Darwin that the low birth rate of the professional and upper class entailed a deterioration of the genetic quality of the population.

5. KARL PEARSON RESTATES THE PROBLEM

In the next generation Karl Pearson (1857–1936) was the leading exponent of the theory of genetic deterioration. Pearson studied mathematics at Cambridge, became professor of mathematics at University College, London, and, in 1911, became the first holder of the Galton Professorship of Eugenics, a post founded by a bequest in Galton's will. Pearson's first important contribution was to clarify the problem of regression to the mean. Galton had demonstrated by experiments on peas that long pods produce offspring a little shorter than themselves, while short pods produce pods a little longer. It was realized that this is a general phenomenon of nature and holds for a number of human characteristics including ability. This had also been shown by Galton in his pedigree studies, where he found that the sons of eminent men were not on average as eminent as their fathers. The problem was that if regression to the mean continued to operate over several generations, the descendants of the highly intelligent and of the very dull would regress to the mean of the population. If this was so, it would not matter that the more talented classes had lower fertility than the untalented, since in a few generations their descendants would be indistinguishable from one another. In response to this problem, Pearson (1903) worked out a statistical theory which showed that selective breeding altered the population mean, and it was to the new mean that offspring would regress. This meant that the inverse association between ability and fertility was shifting the population mean of ability downward, and the genetic deterioration entailed by this would not be corrected by regression effects.

Pearson turned his attention next to the issue of the heritability of intelligence. The problem here was that genetic deterioration resulting from the inverse association between ability and fertility would only take place if intelligence has some heritability. Galton appreciated this and set out arguments for a high heritability of intelligence based on pedigree studies showing the transmission of high ability in families, but Pearson realized that the case needed strengthening. He tackled this problem by making a study of the correlation between pairs of siblings for intelligence and for the physical characteristics of height, forearm length and hair and eye color. Data were collected from London schoolchildren for approximately four thousand pairs of siblings; their physical characteristics were measured and their intelligence assessed by teachers' ratings (Pearson, 1903).

To examine the strength of association for these characteristics between the sibling pairs, Pearson worked out the mathematics of the correlation coefficient. He found that all the characteristics were correlated among siblings at a magnitude of approximately 0.5. Numerous subsequent studies using intelligence tests have confirmed this result, the average correlation based on 26,000 pairs being calculated by Bouchard (1993, p. 54) at 0.47. Pearson's argument was that physical characteristics which are obviously under genetic control, like hair and eye color, and stature and forearm length among siblings, are correlated at a magnitude of about 0.5. This is also the case with intelligence and suggests that intelligence is likewise under genetic control. More generally, an environmental theory of the determination of intelligence would predict much more similarity between siblings than is actually present, because siblings are raised in closely similar environments and would be expected to have closely similar levels of intelligence. The relatively low correlation between siblings shows that genetic factors must be operating to differentiate them.

Pearson followed Galton, Darwin and Spencer in believing that natural selection had largely ceased to operate in modern populations as a result of the increased survival and high fertility of the less fit. In his 1901 book, *National Life*, he wrote that "while modern social conditions are removing the crude physical checks which the unrestrained struggle for existence places on the over-fertility of the unfit, they may at the same time be leading to a lessened relative fertility in those physically and mentally fitter stocks, from which the bulk of our leaders in all fields of activity have hitherto been drawn" (p. 101). Eleven years later he reaffirmed that "the less fit were the more fertile" and consequently "the process of deterioration is in progress" (1912, p. 32). Pearson drew the same conclusion as Galton that the only way to counteract genetic deterioration was by eugenic intervention. "The only remedy," he wrote, "if one be possible at all, is to alter the relative fertility of the good and the bad stocks in the community" (1903, p. 233).

6. THE RISE OF EUGENICS

In the early decades of the twentieth century increasing number of biological and social scientists accepted the thesis that natural selection against the less fit had largely ceased to operate in modern populations, that those with high intelligence and strong character had begun to have lower than average fertility and that this must cause a genetic deterioration of the population. From this set of premises, many people were drawn irresistibly to the logic of Francis Galton that the only way to avert genetic deterioration was by finding consciously designed measures to replace what had formerly been done by natural selection. Those who were concerned with this question founded eugenics societies whose general purpose was to carry out research on the issues of heritability and genetic deterioration, to inform public opinion of the seriousness of the problem and to formulate policy proposals to counteract deterioration and replace it by improvement. In 1906 the American Breeders' Association, renamed the American Genetics Association in 1913, set up a Committee on Eugenics to promote the work on the concept, and in 1910 the Eugenics Record Office was established at Cold Spring Harbor on Long Island as a center for eugenic research and publication. The American Eugenics Society was formed in 1923. Eugenics societies were established in Germany in 1906 and in Britain in 1907, and by 1930 eugenics societies had been set up in many other countries, including Latin America, Australia, Canada, Japan and virtually the whole of Continental Europe.

Most of the leading biological and social scientists in the first half of the twentieth century were members of these societies and subscribed to their objectives. In Britain they included the biologists and geneticists Sir Ronald Fisher, Sir Julian Huxley and J.B.S. Haldane and the psychologists Sir Cyril Burt, Sir Godfrey Thomson and Raymond Cattell, who did most of the early work on the question of whether intelligence is declining. In the United States they included the geneticists Hermann Muller and Charles Davenport, who discovered that Huntington's Chorea is inherited by a single dominant gene, and the psychologists Robert Yerkes, who constructed the Army Alpha and Beta intelligence tests, and Lewis Terman, who set up the study of approximately 1,500 highly intelligent children who have been followed up over their life span. Although there were some dissenters, in the first half of the twentieth century, many of the leading biologists and social scientists accepted that modern populations were undergoing genetic deterioration and that eugenic measures needed to be found to correct this.

7. RONALD FISHER ON THE DECAY OF CIVILIZATIONS

Among geneticists and biologists of the middle decades of the twentieth century who believed that natural selection had broken down, that genetic deterioration was taking place and that eugenic measures needed to be designed to counteract this, the foremost were Ronald Fisher, Julian Huxley and Hermann Muller.

Ronald Fisher (1890–1962) was both a geneticist and a statistician. He graduated in mathematics at Cambridge and worked initially at the Rothamsted Experimental Station for genetic research. There he developed the mathematics of the statistical method of analysis of variance and, as his biographers say, "recast the whole theoretical basis of mathematical statistics and developed the modern techniques of the design and analysis of experiments" (Yates and Mather, 1963, p. 92). In 1933 Fisher was appointed as Karl Pearson's successor to the Galton Professorship of Eugenics at University College, London, and in 1943 he became professor of genetics at Cambridge where he remained until his retirement.

Fisher was twenty-eight when he published his first important paper on genetics integrating Mendelian single gene and polygenetic (multiple gene) processes (Fisher, 1918). Up to this time the dominant paradigm in genetics was Mendel's theory of the action of single genes, the effect of which was to produce two types of individuals. It was obvious that for many traits, such as height, skin color, intelligence and so on, there are not only two types, but a continuous gradation. Fisher demonstrated mathematically that traits of this kind could be explained by the joint action of a number of genes acting according to Mendelian principles, each of which contributed equally and additively to the determination of the trait. He showed that this would lead to correlations of about .5 between parents and children and between pairs of siblings; that it was possible to partition the variance of the trait into heritable and environmentally determined fractions; that the heritable fraction could be divided into that caused by additive genes, by dominance and genetic interaction; and that the correlation between siblings was further affected by ''assortive mating,'' the tendency of people to mate with those like themselves, which raises the correlation between their children. This path-breaking paper laid the foundations of what was to become the science of biometrical genetics.

Fisher took up the issues of the breakdown of natural selection and genetic deterioration in his 1929 book *The Genetical Theory of Natural Selection.* In this he summarized a number of the early investigations showing an inverse association between socioeconomic status and fertility. For instance, Heron (1906) had shown that among London boroughs the birth rate was inversely associated with an index of average socioeconomic status. The 1911 census for England was analyzed by Stevenson (1920), who estimated that the fertility of the social classes was lowest among the professional and upper class, who had an average of 1.68 children, and increased progressively among the lower middle class (2.05), skilled workers (2.32), semiskilled (2.37) and unskilled (2.68). The census of 1906 in France showed a similar trend with the average number of children of the middle class being 3.00 and of the working class, 4.04.

Fisher followed Galton in believing that social mobility over the course of centuries had led to the disproportionate concentration of the genes for high intelligence and strong work motivation in the professional class, and that, as a consequence, the low fertility of the professional class must entail genetic deterioration of the population in respect of these qualities. Fisher also followed and elaborated on Galton in his explanation for the inverse relationship between socioeconomic status and fertility: This was that intelligent and well-motivated young men rise in the social hierarchy and tend to marry heiresses as a way of consolidating their social position. Heiresses tend to come from relatively infertile stocks, because if the stocks had high fertility these women would have had brothers and would not be heiresses. The effect of this was that able men tended to marry infertile women, and so had few children. He cited data in support of his contention that fertility does have some heritability. Fisher proposed that this process has frequently occurred in the history of civilizations and explained their decay, and he instanced classical Greece, Rome and Islam as examples. He proposed a universal sociological law asserting that advanced civilizations are characterized by dysgenic fertility, and that this leads to genetic deterioration and ultimately to the decay of civilization.

The Galton-Fisher theory of the causes of dysgenic fertility is rather implausible for several reasons. First, it is doubtful whether such an important fitness characteristic as fertility has any significant heritability because individuals who carried the genes for low fertility would have left fewer descendants, and these genes would have been eliminated. Second, even if fertility does have some heritability, it is questionable whether many able young men are sufficiently calculating to seek out and marry heiresses. Both Galton and Fisher forgot about the power of love in the selection of marriage partners. Fisher certainly did not follow his own theory because he married a fertile nonheiress by whom he had eight children. Third, there is no strong evidence to support the thesis that the decay of past civilizations has been due to dysgenic fertility. Fourth, the inverse association between socioeconomic status and fertility in modern populations is most plausibly explained by the efficient use of contraception by the professional and middle classes and the increasingly less efficient use of it by the working classes. There is a class gradient for intelligence and the personality qualities of restraint, farsightedness and the capacity to "delay gratification," as we shall see in detail in Chapters 11, 12 and 14, and it is the social gradient of these qualities that is the principal cause of dysgenic fertility.

In spite of these criticisms of the Galton-Fisher theory of the causes of dysgenic fertility, we should not lose sight of the main point that Fisher believed there is a genetically based social-class gradient for intelligence and socially valuable personality traits, and hence that the inverse association between socioeconomic status and fertility entails genetic deterioration. Fisher saw this biological success of the social failures as the greatest threat to the future of our civilization.

8. JULIAN HUXLEY AND THE EVOLUTIONARY PERSPECTIVE

Julian Huxley (1887–1975) was the grandson of T.H. Huxley, the Victorian biologist who was known as Darwin's bulldog because of his rigorous defense of the theory of evolution, and the brother of Aldous Huxley who described a eugenic state in his novel *Brave New World.* Huxley studied biology at Oxford and went on to become chairman of the genetics department at the Rice Institute in Texas; professor of physiology at King's College, London; secretary to the London Zoological Society and, in 1946, the first Director General of UNESCO (The United Nations Educational, Scientific, and Cultural Organization). Huxley was primarily an evolutionary biologist whose major work was the integration in his book *Evolution, the Modern Synthesis* (1942) of Darwinian theory with Mendelian genetics.

Huxley was president of the British Eugenics Society from 1959 to 1962. He set out his views on eugenics in two lectures delivered to the society, the first in 1936 and the second in 1962. In the first he restated the argument that genetic deterioration is taking place in modern populations as a result of the relaxation of natural selection and of the inverse relationship between social class and fertility:

> Deleterious mutations far outnumber useful ones. There is an inherent tendency for the hereditary constitution to degrade itself. [But] in wild animals and plants, this tendency is either reversed or at least held in check by the operation of natural selection, [and] in domestic animals and plants, the same result is achieved by our artificial selection. But in civilised human communities of our present type, the elimination of defect by natural selection is largely (though of course by no means wholly) rendered inoperative by medicine, charity, and the social services; while, as we have seen, there is no selection encouraging favourable variations. The net result is that many deleterious mutations can and do survive, and the tendency to degradation of the gene-plasm can manifest itself. Today, thanks to the last fifteen years' work in pure science, we can be sure of this alarming fact, whereas previously it was only a vague surmise. Humanity will gradually destroy itself from within, will decay in its very core and essence, if this slow but insidious relentless process is not checked. Here again, dealing with defectives in the present system can be at best a palliative. We must be able to pick out the genetically inferior stocks with more certainty, and we must set in motion counter-forces making for faster reproduction of superior stocks, if we are to reverse or even arrest the trend. (1936, p. 30)

Huxley believed that the two major causes of genetic deterioration in modern populations were the growth of social welfare, which was undermining the operation of natural selection in eliminating defective stocks by high mortality, and the development of the inverse relation between socioeconomic status and fertility. Like Galton, Pearson and Fisher, he believed that the professional class is a genetic elite with regard to intelligence, and its low fertility must inevitably lead to genetic decline.

In his 1962 lecture Huxley reasserted and elaborated on these points. He argued that those with genetic defects and what had come to be known as "the social problem group" should be discouraged from having children. He suggested that this could be achieved by voluntary sterilization and instruction and assistance in the efficient use of birth control.

9. HERMANN MULLER'S GENETICISTS' MANIFESTO

Hermann Muller (1891–1967) recorded that he first became interested in eugenics at the age of ten when his father took him to the New York Museum of Natural History and explained the display of the evolution of horses' hooves. If horses' hooves could be improved by unplanned natural selection, the precocious child wondered, could not human beings also be improved by planned selection. Muller studied biology and genetics at Columbia University. In his first year he presented a paper on eugenics at a meeting of one of the college societies. In this he noted the relaxation of natural selection in contemporary societies, that this would result in a failure to eliminate detrimental mutations and that these mutants would accumulate, causing an increase in what he called the "genetic load."

Muller spent the early years of his career as a research geneticist working on

Drosophila, in particular on the effects of radiation in increasing mutations. He was in his mid-forties when he published his first major work on eugenics, *Out of the Night* (1935). In this he restated the mainline eugenic thesis that modern populations were deteriorating in regard to health, intelligence and character because of the reduction of mortality and dysgenic fertility: "The more shiftless, the less intelligent and the less progressive members of our communities are reproducing at a higher rate than those of a more desirable type." Muller was the first advocate of the establishment of elite sperm banks as a possible way of reversing genetic deterioration. His idea was that "the greatest living men of mind, body, or spirit" would deposit sperm in these banks, and that many ordinary women should be encouraged to make use of this by artificial insemination to produce children of high genetic quality. The donors would be men of sound health, high intelligence and also of co-operative and altruistic character.

In 1939 Muller attended the International Congress of Genetics in Edinburgh and while he was there drew up a document called "The Geneticists' Manifesto." This addressed the question, How could the world's population be improved most effectively genetically? The answers given to this question were, first, that the environment needed to be improved and made more egalitarian to allow those classes who were handicapped by impoverished social conditions to realize their full genetic potential. Second, it was stated that "the intrinsic (genetic) characteristics of any generation can be better than those of the preceding generation only as a result of some kind of selection, i.e. by those persons of the preceding generation who had better genetic equipment having produced more offspring, on the whole, than the rest." It was then stated that "under modern civilised conditions such selection is far less likely to be automatic than under primitive conditions" and hence "some kind of conscious guidance of selection is called for." This guidance should take the form of measures to increase the fertility of those with the qualities of health, intelligence and "those temperamental qualities which favour fellow-feeling and social behavior" (Muller et al., 1939, p. 64).

Muller got a number of the leading geneticists of the day attending the conference to sign this manifesto, including J.B.S. Haldane, S.C. Harland, L. Hogben, J. Huxley and J. Needham and it was subsequently endorsed by a number of others. In the late 1930s there was a wide measure of consensus among geneticists that natural selection was no longer working effectively and that consciously designed selection would have to be introduced to prevent genetic deterioration.

10. RAYMOND CATTELL AND THE DECLINE OF INTELLIGENCE

In the 1920s and 1930s the issue of whether intelligence was deteriorating was taken up by a number of psychologists in the United States and Britain.

Hitherto people like Galton, Pearson, Fisher and Huxley had relied on the growing evidence that there was an inverse relationship between social class and fertility. They assumed that there was a positive relationship between social class and intelligence and argued that this implied that the intelligence of the population must be declining. With the invention of the intelligence test by Alfred Binet in France in 1905, it became possible for psychologists to carry out studies to determine whether this was actually occurring.

The psychologist who did the most work on this issue was Raymond Cattell, a British psychologist who graduated in chemistry at the University of London, and then switched to psychology to do postgraduate work on intelligence with Charles Spearman. Cattell worked on the question of declining intelligence in the 1930s and set out his results in his book *The Fight for our National Intelligence* (1937). In this he made five principal contributions to the problem. First, he constructed and used a new type of "culture fair" intelligence test. Up to this time intelligence tests had been largely composed of vocabulary and general knowledge questions, and critics were able to argue that performance on these simply reflected differences in education and social class rather than innate ability. Cattell's culture fair test consisted of problems in design and pictorial format, such as a randomly arranged set of pictures from a strip cartoon which had to be put into the correct temporal sequence. Cattell's new culture fair test measured what he was later to call "fluid intelligence" and improved the credibility of the test.

Second, Cattell collected normative data which showed the existence of a social class gradient for intelligence. According to his results, the mean IQ in the higher professional class was in the range 142–151; among minor professional and other white collar occupations, 115–137; among skilled workers, 97–114; among the semiskilled, 87–96; and among the unskilled, 78. This was one of the first studies to show that the assumption that the social classes differ in intelligence was correct. The range of these differences is rather greater than would be expected. The explanation for this is that Cattell's test had a standard deviation of 24 rather than fifteen, which later came to be adopted.

Third, Cattell made an estimate of the decline of intelligence. The method he used was to collect data on the intelligence of a sample of approximately 3,700 ten-year-old children and examine this in relation to their number of siblings. The result showed that the more intelligent the children, the fewer their average number of siblings, and from this Cattell inferred that the intelligence of the population must be declining. He calculated the rate of decline at approximately three IQ points a generation. Later studies in Britain by Sir Cyril Burt (1946) and Sir Godfrey Thomson (1946) reached similar conclusions.

Cattell went on to set out the consequences of the deterioration of intelligence for the quality of national life. He predicted a decline in educational attainment, in moral standards, and in cultural, scientific and economic life and in law-abiding behavior. Finally, he made some suggestions for eugenic measures to arrest the decline of intelligence, the most important of which were the provision of financial incentives for the more intelligent to have children and the more effective provision of birth control facilities for the less able.

A little over a decade later Cattell carried out another study on a comparable sample of ten-year-olds to see whether their average intelligence had declined. The result was that the average IQ had increased by 1.3 IQ points (1951). This was to be a common finding of a number of other studies and has come to be known as "the Cattell paradox." Cattell himself proposed that the explanation was that various environmental improvements, especially those in education, had masked a real decline. We shall be examining these questions in detail in Chapters 5 through 8.

11. THE 1963 CIBA CONFERENCE

In the second half of the twentieth century, the theory that modern populations are deteriorating genetically and the associated belief that eugenic measures need to be taken to correct this began to lose ground. Nevertheless, prominent biologists and social scientists were still voicing eugenic concerns. In 1963 the Ciba Foundation convened a conference in London on the theme of *Man and His Future*. The conference included a session on eugenics at which papers were delivered by Hermann Muller and Joshua Lederberg, followed by a discussion. Muller, at this time a geneticist at the University of Indiana, restated his earlier views that genetic deterioration was taking place through the accumulation of harmful genes, which were ceasing to be eliminated by natural selection, and as a result of the inverse association between social class and fertility, which must be dysgenic because of the genetic superiority of the higher social classes. To counteract genetic deterioration he advocated the establishment of elite sperm banks which women wishing to have children would be encouraged to use, although several of the discussants expressed doubt about whether this facility would be widely taken up.

The second paper was given by Joshua Lederberg, professor of genetics and head of the medical school at Stanford. He began by restating the problem of genetic deterioration: "Most geneticists . . . are deeply concerned over the status and prospects of the human genotype. Human talents are widely disparate; much of the disparity has a genetic basis. The facts of human reproduction are all gloomy—the stratification of fecundity by economic status, the new environmental insults to our genes, the sheltering by humanitarian medicine of once lethal defects" (p. 264). He went on to discuss how genetic deterioration in Western nations could be overcome and proposed that genetic engineering might offer a better approach than attempting to alter fertility rates in a more eugenic direction.

In the discussion that followed Julian Huxley was strongly supportive of the eugenic position. So also was Francis Crick, who began by saying that he agreed with the general eugenic case that genetic deterioration was taking place and that "I think that we would all agree that on a long-term basis we have to do something.'' He proposed that it was time to challenge the belief that everyone has a right to have children, and suggested that a system of licensing might be introduced under which only those with socially desirable qualities would be permitted to reproduce. Alternatively, he suggested that because, by and large, people with high incomes had more socially desirable qualities, particularly those of industry, than the poor, a tax on children might be imposed which would act as a deterrent on the poor from reproducing but would have little effect on the financially better off (p. 274).

The Ciba Foundation conference of 1963 was to prove the high water mark of eugenics. Three of the most distinguished biologists of the time, Hermann Muller, Joshua Lederberg and Francis Crick, all of whom had won the Nobel Prize for their work on genetics, recognized the seriousness of the problem of genetic deterioration in modern civilizations and proposed methods for counteracting it. It was a high water mark from which the tide was to ebb rapidly. Over the next quarter of a century support for eugenics was to evaporate. By 1985 Daniel Kevles, a leading historian of the eugenics movement, pronounced that ''eugenics is dead,'' and he was right.

12. THE DECLINE AND FALL OF EUGENICS

Throughout the twentieth century there had always been some who dissented from the eugenicists' concerns about the genetic deterioration of modern populations. In the 1930s Launcelot Hogben (1931) attacked eugenics for being based on social class and race prejudice. In the post–World War II years, Lionel Penrose was another prominent opponent. In 1945 Penrose was appointed to Galton Professorship of Eugenics at University College, London. He rapidly had the name changed to the Galton Professorship of Genetics, disputed the thesis that the intelligence of the population was deteriorating and denounced the ''cult of eugenics'' (1948, 1967). After 1945, the misuse of eugenics by the Nazis, in their mass sterilization program and the extermination of the Jews and gypsies, was used to discredit eugenics by people who maintained that eugenics inevitably led to the gas chambers (Kaye, 1987).

From the 1960s onward, eugenics became increasingly repudiated. The officers of the eugenics societies in the United States and Britain lost their nerve, changed the names of the societies and closed down their journals to dissociate themselves from the concept. Occasionally the theory of genetic deterioration and the need for corrective eugenic measures was restated, notably by Robert Graham (1970), William Shockley (1974) and Raymond Cattell (1972, 1987), but their work attracted little attention and virtually no favorable comment. Eugenics was attacked in three histories of the movement by Mark Haller (1963), Kenneth Ludmerer (1972) and Daniel Kevles (1985), all of whom argued that the whole eugenic case was misconceived, that the eugenicists did not understand genetics, that there were no genetic differences between the social classes, that the low fertility of the professional class was of no consequence and that no genetic deterioration was taking place. Kevles, in the most influential of these histories, *In the Name of Eugenics*, represented eugenics as a kind of crackpot religion in which Galton was "the Founder of the Faith," Karl Pearson was dubbed "Saint Biometrika," and Charles Davenport, the leading American eugenicist of the early twentieth century, became a "Worshipper of Great Concepts." Kevles dismissed the growing acceptance of eugenics in the first half of the century with the phrase "the gospel becomes popular" and lost no opportunity to sneer at the eugenicists for being "obsessed by the procreational practices of others" (p. 286). Eugenics, according to Kevles, was based on "false biology," and there was no need for concern about a negative association between intelligence and fertility because hereditarian theories of intelligence had been disproved by two people, Brian Evans and Bernard Waites. Nowhere did Kevles show any comprehension of the genetic problems entailed by the breakdown of natural selection in Western populations, of the evidence pointing to genetic deterioration, the seriousness of this issue, or the validity of the concerns felt by eugenicists and their attempts to find solutions. Half a century earlier, Kevles' book would have been trashed for missing all the major points of the eugenic argument, but by the 1980s it was uncritically applauded by a sympathetic media. Extracts from the book were serialized in the *New Yorker*, and the *New York Times Book Review* described it as "a revealing study by a distinguished historian of science."

By the last decade of the century eugenics had become universally rejected. In 1990 University College, London, appointed Steve Jones as head of the Galton Laboratory. Jones's expertise lay in the genetics of snails. He had no sympathy whatever with eugenics, and his view of Galton was that he was "a fascist swine" (Grove, 1991). Four years later a leading British geneticist, Sir Walter Bodmer, was to write of "the mindless practice of eugenics" (Bodmer and McKie, 1994, p. 236).

By the 1990s even the members of the former eugenics societies attacked eugenics. In 1991 the British Eugenics Society, by now blandly renamed *The Galton Institute*, devoted its annual conference to Francis Galton and his achievements, and the conference proceedings were later published, edited by Milo Keynes (1993). The conference consisted of lectures on Galton's various interests including travel, the weather, statistics, physical growth, fingerprints and education. Strangely absent was any lecture on eugenics, Galton's principal interest over a period of almost half a century. No mention whatever was made of Galton's views on the problem of genetic deterioration among Western populations and only a brief and dismissive reference was made by Keynes to Galton's ideas on eugenics. Keynes denounced Galton's ideas on the grounds that there are no genetic elites and therefore there could be no genetic gains in encouraging them to increase their fertility. These ideas, according to Keynes, were all wrong because Galton did not know of the work of Mendel: "Through his ignorance of Mendelism, Galton was unbiological when he thought that eugenic policies could be achieved by encouraging the fertility of families in which eminent men belonged'' (1993, p. 23). As a matter of fact, Galton was not ignorant of Mendel's work and he paid tribute to it in his *Memoirs* (1909, p. 308). Furthermore, it is not necessary to know the details of Mendelian gene processes in order to breed better stocks. This has been done by animal breeders for centuries, as Galton pointed out on the first page of his *Hereditary Genius.* But by the 1990s this was no longer understood by the leading members of the former British Eugenics Society. The collapse of eugenics was complete.

13. CONCLUSIONS

In this chapter we have seen how in the second half of the nineteenth century Benedict Morel in France, and Francis Galton, Charles Darwin and Herbert Spencer in England realized that natural selection had largely ceased to operate in Western populations and that the result of this would be genetic deterioration. They understood that this deterioration was taking place for health, intelligence and character, and that the reason for this was that those who were deficient in these regards, who had previously suffered high mortality and had low fertility, now had reduced mortality and, in the case of intelligence and character, had begun to have high fertility. These Victorian biologists and social scientists perceived that the genetic deterioration resulting from the relaxation of natural selection could only be countered by some form of consciously planned selection. Francis Galton proposed the term *eugenics* for the study of this problem. In the early and middle decades of the twentieth century these ideas came to be accepted by many leading biological and social scientists, including Karl Pearson, Charles Davenport, Sir Ronald Fisher, Sir Julian Huxley, Sir Cyril Burt, Sir Godfrey Thomson, Raymond Cattell, Hermann Muller, Joshua Lederberg and Francis Crick.

From the 1960s a reaction against eugenics set in and by the last decade of the century the concept was virtually universally rejected. My objectives in this book are to show that the eugenicists were right in their belief that natural selection has broken down and that, as a consequence, genetic deterioration is occurring in modern populations; to present the evidence for this; and to assess the magnitude of the problem. It may be helpful to set out the framework within which this task is attempted. We look first at the root cause of the problem, the way in which natural selection preserved the genetic quality of the population in preindustrial societies (Chapter 2) and how natural selection has broken down in the nineteenth and twentieth centuries (Chapter 3). We look next at this breakdown in regard to health (Chapter 4), intelligence (Chapters 5 through 11) and character (Chapters 12 through 14), and then at the issue of genetic deterioration in economically less developed nations (Chapter 15). Finally (Chapter 16), we consider the arguments against the view that Western populations are undergoing genetic deterioration, find them wanting and conclude that the eugenicists were right in identifying this as a serious problem, and one that needs attention.

Chapter 2

Natural Selection in Preindustrial Societies

1. General Principles of Natural Selection. 2. Natural Selection against Disease. 3. Hunter-Gatherers. 4. Pastoralists and Agriculturalists. 5. Early Nation-States. 6. Christendom. 7. Fertility in Europe, 1500–1850. 8. Socioeconomic Differentials in Reproductive Fitness. 9. Social Mobility in Preindustrial Societies. 10. High Mortality of Illegitimates. 11. Conclusions.

The root cause of the genetic deterioration of the populations of the Western nations is that natural selection against those with poor health became relaxed; and natural selection against those with low intelligence and poor character ceased to operate during the course of the nineteenth century. To understand the significance of this development, we need to consider how natural selection worked to keep the populations genetically sound in preindustrial human societies, and the dramatic nature of the change that came about when natural selection broke down. This is our concern in the present chapter.

1. GENERAL PRINCIPLES OF NATURAL SELECTION

There are two processes by which natural selection keeps a population genetically sound. These are through the greater tendency of those with genetically superior qualities to survive—"the survival of the fittest"—and through their higher fertility. Conversely, those with genetically inferior qualities have greater mortality and lower fertility. These two processes apply quite generally for health throughout all animal species. Individuals who lack genetic immunities or who inherit genetic disorders tend not to reproduce—either because they are infertile, or because they are physically impaired, or because they make unattractive mates—and to die in infancy or childhood before they have an opportunity to reproduce. Through these mechanisms the genes responsible for genetic disorders and diseases are constantly eliminated from the population.

With regard to intelligence and what the eugenicists called character, natural selection in preindustrial human societies operated principally through males. It is a general principle throughout animal species that males compete with one another for females. Typically it is the stronger, more aggressive and healthier males who succeed in these competitions, secure females and consequently transmit their genes for these characteristics to their progeny. Competition between males is generally reinforced by a biologically programmed propensity of females to prefer successful males as mating partners.

These principles of natural selection were understood and set out by Darwin in *The Descent of Man*, where he wrote, "It is certain that amongst all animals there is a struggle between the males for possession of the female" (1871). Subsequent observations by naturalists were to show that this was right and were synthesized by the Scottish biologist Wynne-Edwards (1962), who showed that throughout animal species, males compete either for territory or status and that only those who are successful gain access to females and are able to reproduce. In a number of species of fish, amphibia and birds, competition between males is largely for territory. In any particular locality there are only a certain number of territories, and this limits the number of males able to obtain them. The males who secure territories are the strongest, the most aggressive and the healthiest. Females will only mate with males who have obtained territories, and hence only the fittest reproduce.

Most mammals are social animals who live in groups, and males do not have personal territories. In these species the males are typically ordered in a status or dominance hierarchy, and only the high-ranking ones have access to females. As young males come into adulthood, they compete to secure acceptance as members of the group and, once this is achieved, to gain high status in the hierarchy. Those who are the strongest, most aggressive and healthy are successful and able to secure mates.

As human beings evolved from apes over the course of the last five million years or so, males came to need intelligence and "character" to be successful in competition for status. We can envisage what this competition must have been like among evolving humans by looking at the social organization of troupes of apes and monkeys and of primitive peoples, and where these are similar it is likely that the same processes were present in the evolution of humans.

Typically apes, monkeys and human hunter-gatherers live in groups of around 20 to 50 individuals. The males are ordered in status hierarchies in which the higher-ranking males have greater access to females. For longish periods of time there is no movement in the hierarchy. Then the male at the top falls ill, weakens or dies, and this opens up an opportunity for middle-ranking males to challenge and replace him. The young male seeking to rise in the hierarchy has to make fine judgments and exercise restraint about when to accept his place in the hierarchy and when to challenge it. Those who are too aggressive at too young an age are at risk of being killed or expelled from the troupe, while those who are too timid fail to move up the hierarchy when niches in the middle and at the top fall vacant. Just how to behave in order to secure an improved position in the hierarchy and eventually reach the top requires both intelligence and "character," that syndrome of personality qualities comprising self-discipline, restraint, the capacity to work steadily over a period of many years for long-term goals; the ability to cooperate with others and form political alliances; and the integrity required to gain the approval of colleagues and superiors. Over the period during which humans evolved, males possessing intelligence and character were more successful in competition for status and, as a consequence, access to females, and the genes responsible for these qualities gradually evolved and spread in human populations.

2. NATURAL SELECTION AGAINST DISEASE

Throughout the animal kingdom and in human societies from the first hunter-gatherers to Western societies up to the nineteenth century, natural selection operated against individuals with poor health and with genetic diseases. Generally most women had at least four children and frequently many more, but the populations grew only slowly because about half of those born died before they reached adulthood. Those who died young were predominantly those who lacked the genetic immunities against disease, who had poor nutrition which lowered their resistance to disease, and who had genetic diseases that were incurable. High infant and child mortality kept the population genetically sound in regard to health up to around the year 1800, after which improvements in sanitation and other medical advances led to a reduction in mortality and weakened the impact of natural selection against disease.

3. HUNTER-GATHERERS

In addition to natural selection against those with poor health, natural selection has also operated in favor of intelligence and character among the hunter-gatherer peoples who still survive in remote parts of the world and whose life-styles have been described by anthropologists. These people live in small groups, like most other primates. Generally about half to three quarters of the infants born die before they reach adulthood as a result of disease, accidents or warfare in which the victors kill the defeated.

The males in hunter-gatherer peoples are ordered in status hierarchies, as in many mammalian species, and high-status males have greater reproductive success. This was understood by Darwin who described it as follows: "The strongest and most vigorous men—those who could best defend and hunt for their families, and during later times the chiefs or head-men . . . would have succeeded in leaving a greater average number of offspring than would the weaker, poorer and lower members of the same tribes. . . . The chiefs of nearly every tribe throughout the world succeed in obtaining more than one wife'' (1871, Vol. 2, p. 368). Subsequent research by anthropologists and biologists has shown that this was correct. Typically, the males in hunter-gatherer societies fall into three bands. First, there is the leader or headman. who has several wives or access to several women, who may include the wives of others. Second, there are males of intermediate status who have one wife; finally, there are low-status males who have no wives. The result of these social systems is that the leaders have the most children; the intermediate males have a few, while the low-ranking males have none. Hence, the qualities necessary to secure group leadership are genetically enhanced, and these include health, intelligence and character.

A good description of a hunter-gatherer people with this lifestyle has been given by Neel (1983) and Chagnon (1983) of the American Indian Yanomamo of the upper Amazon basin. These peoples live in small groups led by a headman, who has two or three wives, while the other men have one or none. Among a number of these groups, Neel and Chagnon found that headmen had an average of 8.6 living children. To become a headman requires intelligence, including good verbal and reasoning abilities to out-talk others and good spatial abilities for effective hunting. As David Buss in a review of studies of these groups observes: ''In tribal societies the headmen or leaders are inevitably among the most intelligent in the group'' (1994, p. 34). And success in competition to become a headman also requires the character qualities of self-discipline needed to command the respect and loyalty of the group.

Among these hunter-gatherer peoples, there are incessant conflicts between neighboring groups in which the victors generally kill off the defeated men and boys and take over the women. The groups that succeed in these conflicts are those with better health, physical strength, intelligence and character, such as the ability to plan ahead, control impulsiveness and exercise caution. Thus, natural selection operates to strengthen these qualities through conflicts between groups, as well as through competition between males for status within groups. A similar social system has been described by Hill and Kaplan (1988) among the Ache, an American Indian hunter-gatherer people in Paraguay.

Another example of a contemporary hunter-gatherer people with this lifestyle is the !Kung San tribes of the Kalahari desert. They live in small bands and move their camps frequently from one watering hole to another. Howell (1979) has made a study of them and estimates that 62 percent of the males produce no children. About 45 percent of children die before they reach adulthood. Mortality is higher among males than among females, so there is an excess of females, and this allows about 5 percent of males to have two or more wives.

The lifestyle of hunter-gatherers has been reviewed by Murdock (1967) and Betzig (1986), who have shown that the great majority of these people have status-based polygamous mating systems of this kind. Natural selection operates by differential mortality and by differential fertility in favor of males who are successful in competition for status. This ensures greater reproductive success for males who are healthier, more intelligent and have stronger character.

4. PASTORALISTS AND AGRICULTURALISTS

A number of hunter-gatherer peoples have evolved into migratory pastoralists and settled agriculturalists. Where this has occurred they continue to practice polygamy. Pastoralists are nomadic peoples who keep domesticated animals, such as goats and camels, and migrate from place to place to find new pastures for the herds. Their diet consists largely of produce from these herds, mainly cheese, milk and blood, and from time to time the animals which they slaughter for meat. They often trade their animal products for plant foods obtained from settled agricultural peoples. The first pastoralists appeared in Neolithic times, about ten thousand years ago around the shores of the Mediterranean, in the Middle East, Asia and East Africa (Harris and Ross, 1987).

The size of the population that pastoralists can sustain is constrained by the size of their herds. A good description of a typical pastoral people, the Rendille camel herders of Northern Kenya, is given by Moran (1979). These people enforce celibacy on all males up to the age of 31. Once they have attained this age males are permitted to marry, but to do this they have to produce a brideprice which is paid in camels. Only about 50 percent of the males are able to accumulate sufficient camels for the required brideprice, and those who are unable to do this remain wifeless and childless. Those who are able to accumulate the required number of camels are allowed to buy several wives, if they can afford them. The key to reproduction is, therefore, the ability to accumulate wealth in the form of camels. There can be little doubt that intelligence, character, the ability to exercise restraint and to work for long-term objectives make an important contribution to the acquisition of sufficient camels to buy a wife and have children.

A number of pastoral people have settled down to become agriculturalists. Typical peoples of this kind, described by anthropologists, are the Yomut tribesmen in Iran, described by Irons (1979), and tribal peoples in Nigeria (Driesen, 1972), New Guinea (Wood, Johnson and Campbell, 1985), Bangladesh (Shaikh and Becker, 1985), and the Kipsigis in Kenya (Borgerhoff Mulder, 1988). Descriptions of over a hundred of these societies have been examined by Betzig (1986) who concludes that virtually all of them have polygamous mating systems in which the successful men have several wives, and therefore greater fertility. In so far as intelligence and character contribute to the achievement of social status in human societies, natural selection for these qualities operates among pastoralists and agriculturalists.

5. EARLY NATION-STATES

About 5,000 years ago, the development of agriculture and the domestication of animals became sufficiently advanced to sustain the first nation-states with populous capital cities and surrounding territories. These nation-states arose initially in river valleys where flood plains produced fertile land in which sufficient crops could be grown to feed the urban populations. The first of these states arose in the valleys of the Euphrates, Tigris, Indus, Nile, Ganges, Yangtze and Hwa Ho rivers.

The rise of the nation-state saw the development of a social-class system. At the top were the ruling families and the heads of the military, police, religious and administrative structures. Below these was a middle class of officers, administrators and merchants; next came skilled artisans, and then laborers, domestic servants, soldiers and slaves. These early nation-states had polygamous marriage systems of a nature broadly similar to those of the agriculturalist societies from which they evolved. High-ranking males typically had numerous wives or concubines. The Old Testament records that King David had more than a hundred of them, while in Turkey and India royal harems sometimes contained over a thousand women. Among the Incas of Peru polygamy was regulated by law according to rank. The emperors had as many wives or concubines as they wanted. Military chiefs were allowed 30 wives and middle-ranking officers were permitted fifteen, eight or seven according to their rank (Betzig, 1986). In China emperors had many hundreds of women in their harems, whom they systematically serviced on a rotating basis at appropriate times in their menstrual cycles, carefully organized and regulated by female supervisors, and through this system they were able to father several hundred offspring (Dickenman, 1979). The all-time record of reproductive fitness for high-status males is believed to be held by Moulay Ismail the Bloodthirsty, a Moroccan emperor who is said to have fathered 888 children (Daly and Wilson, 1983).

In addition to differential fertility in accordance with rank, a second mechanism by which natural selection in early nation-states secured the survival of the fittest was through the better nutrition of the higher social classes. It has been shown in several studies (by measuring the lengths of skeletons in high- and low-class graves) that the higher social classes were better nourished. An example of this comes from the Maya, the extinct civilization of Central America, where skeletons in high-class graves average 7 cm. longer than those in low-class graves (Haviland, 1967). The effects of poor nutrition among the lower social classes would have been to delay menarche in adolescent girls, reduce fertility, impair health and resistance to disease and thereby increase mortality (Frisch, 1978; Frisancho, Matos and Flegel, 1983; Harris and Ross, 1987). The net effect of these factors would have been a lower rate of reproduction in the lower social classes as compared with the higher.

6. CHRISTENDOM

Christianity was adopted as the official religion of the Roman Empire in the year 314. One of its requirements was monogamy. Just why the Roman emperors and their supporting oligarchy should have wished to institutionalize monogamy is rather curious, because polygamy appears better designed to serve the interests of powerful men. Possibly the advantage of monogamy was that it preserved property rights through inheritance for a small number of legitimate children. The support of ordinary Christians for monogamy is easier to understand. Christianity is principally a religion of the poor and the oppressed who receive their reward in heaven for the hardships suffered in their lives on earth, and the poor and oppressed are disadvantaged in polygamous societies because if powerful men have several wives, there are none left for them. So it is in the interests of the poor and the oppressed that women are shared out equally by monogamy.

Whatever the explanation for the institutionalization of monogamy in Christendom, this was the first of the four great blows struck by the Roman Catholic Church against natural selection. The second was the requirement of clerical celibacy which Pope Innocent I imposed about the year 410 and which curtailed the fertility of the many able men and women who entered the church, although it was not always successful in eliminating it entirely. The third was the prohibition of birth control in the nineteenth and twentieth centuries, which made it difficult for the less competent to control their fertility, while the competent found ways of doing so. The fourth was the prohibition of abortion, the availability of which reduces the fertility of the less competent who are more prone to unplanned and unwanted pregnancies.

In spite of the official institutionalization of monogamy, natural selection continued to operate throughout many centuries of Christendom. Powerful men had extramarital relationships which frequently resulted in children. For kings and aristocrats, mistresses were an accepted part of life. For instance, William the Conqueror, who invaded England in 1066, was the illegitimate son of Robert Duke of Normandy and his mistress Arlette. John of Gaunt, Duke of Lancaster (1340–1399) had three illegimate sons. Charles II, King of England in the late seventeenth century, had fourteen illegitimate children, many of whose descendants survive to this day as members of the British aristocracy. In France, Henry II (1499–1566) had Diane de Poitiers as his mistress, Louis XIV (1638–1715) had Madame de Maintenon and Louis XV (1715–1754) had Madame de Pompadour and Madame du Barry.

Even priests and clergy sometimes failed to live up to their vows of celibacy and it was not unknown for cardinals and even popes to have children. In England Cardinal Wolsey had a son, and Pope Alexander VI (1431–1503) had four of them. Occasionally deviant Christian sects persuaded themselves that monogamy was not a requirement of the Christian faith and that God would look favorably on polygamy. One of these was the nineteenth-century American Mormons among whom high-ranking churchmen had an average of five wives and 25 children, while the lower-ranking had one wife and an average of 6.6 children (Daly and Wilson, 1983). So, in spite of the adoption of monogamy, natural selection continued to operate over many centuries in Christendom.

7. FERTILITY IN EUROPE, 1500–1850

From around the year 1500, records of births, deaths and marriages were kept in many towns and villages throughout Europe, and these have been analyzed by historical demographers to ascertain the degree to which people controlled their fertility and whether there were socioeconomic-status differences in mortality, fertility and the number of children surviving to adulthood. As far as the control of fertility is concerned, the evidence for the period 1500 to around 1880 indicates that in general there was no significant control over fertility among those who were married. These had natural fertility, that is to say fertility unrestricted by any methods designed to limit it. The presence of natural fertility can be measured by examination of the degree to which women's fertility declines after the age of 30. When couples limit their fertility, they generally have their desired number of children while the wife is in her twenties and curtail the number born thereafter. This is known as parity-specific fertility because once a specific parity (number of children) has been achieved, couples control further fertility. In conditions of natural fertility wives continue to bear children up to the early to mid-forties. Hence, by examination of the degree to which fertility declines after the age of 30, it is possible to estimate whether fertility is being controlled. This estimate is made by calculation of the statistic m. Where m = 0, fertility is natural; where m is above 0.2, there is some degree of fertility control; and when m = 1.0, fertility control is predominant.

Values of m have been calculated for England from 1550 to 1850 and are virtually zero throughout this period (Wilson, 1982). Data for other countries are summarized by Coale (1986). Zero values for m have been obtained for Germany for the period 1750–1825, for Norway for 1878–1880 and for Sweden for 1871–1880. There is, therefore, widespread evidence that natural fertility was generally present in Europe up to the last quarter of the nineteenth century.

There are, however, some exceptions to this generalization. A number of studies have shown that there was some degree of fertility control among the nobility, the professional classes and the bourgeoisie in several European countries from the late seventeenth century onward. France was unique because natural fertility started to be controlled quite widely in the population from around the year 1800, some eighty years ahead of the rest of Europe. These exceptions constitute the beginnings of the control over fertility, which spread throughout the populations of the economically developed world from around 1880 onward.

Although there was generally prevailing natural fertility among married couples in Europe during the period 1500–1880, apart from the exceptions just noted, the fertility of the total population was kept between four and five children per woman and was therefore significantly below its potential of around eight children per woman. Total fertility rates for a number of historical Western populations have been estimated by Coale (1986). In eighteenth-century Europe these ranged from 4.1 in Denmark and 4.2 in Norway to 4.5 in Sweden and 4.7

in Britain. Substantially higher total fertility rates have been recorded for other populations, such as 6.3 in India for 1906 and 7.0 for the United States in 1800.

Total fertility rates in Europe were kept at moderate levels by three principal means. These were a relatively late age of marriage which reduced the number of children of married women, a high proportion of people who never married and strong social control over extramarital relations among the lower classes. The age of marriage was typically in the late twenties for men and the mid-twenties for women from the sixteenth to the nineteenth centuries. There were strong social pressures preventing people from marrying unless they had a livelihood and a house in which to live and rear a family. Many of those who were unable to find a house and a livelihood found employment as domestic servants. There was an absolute prohibition on these marrying and retaining their positions and as a consequence many of them remained celibate and childless throughout their lives (Wrigley and Schofield, 1981). There were also powerful social pressures against sexual relations outside marriage. The strength of these in New England in the seventeenth century was described by Nathaniel Hawthorne in his novel *The Scarlet Letter*, in which a woman who commits adultery is ostracized and required to wear a shaming red letter ''A'' (adultress) on her dress for the remainder of her life.

In some European countries social ostracism of extramarital sexuality was reinforced by the criminal law. In Britain in the mid–seventeenth century adultery by a married woman was made a capital offense and fornication was punishable by imprisonment, although these activities were decriminalized after the Restoration in 1660 (Beattie, 1986). In Sweden adultery and fornication were punished by flogging, imprisonment and substantial fines from the beginning of the 1600s to the end of the 1800s, and it was not until 1937 that adultery was formally decriminalized. These sanctions were sufficiently powerful to exercise a strong deterrent effect on extramarital sexual activity, to the extent that illegitimacy was kept to approximately 5 per cent of births from the early 1600s through to around 1950 in Britain, Sweden, most of Continental Europe and the United States (Coleman and Salt, 1992; Sundin, 1992; Murray, 1994). Because these social controls operated principally on the lower classes, they had considerable success in checking the fertility of the least competent members of the population.

8. SOCIOECONOMIC DIFFERENTIALS IN REPRODUCTIVE FITNESS

Natural selection worked in favor of intelligence and character in historical times through social-class differences in the number of children born and surviving to adulthood and hence able to have children of their own. Both these factors determine reproductive fitness. We assume for the time being that the upper and middle classes were superior in regard to intelligence and character than the lower classes, as numerous studies in the twentieth century and reviewed in Chapters 11 and 13 have shown. If this was so, the greater reproductive fitness of the upper and middle classes shows the presence of positive natural selection for intelligence and character.

The earliest records showing a lower mortality and higher fertility among higher social classes came from Italy for the fifteenth century. With regard to mortality, Morrison, Kirshner and Molha (1977) have shown that young women in Florence who received large dowries on marriage in the years 1425–1442 lived longer than those who received small dowries. There is similar data for fertility reported for Tuscany for the year 1427 by Herlihy (1965), who has found a positive association between family wealth and the number of children in households. The wealthiest families had more than twice as many living children as the poorest.

Historical demographers have shown that in Europe from the sixteenth to the eighteenth centuries the upper and middle classes had higher fertility and greater number of children surviving to adulthood than the lower classes. Some illustrative statistics from England and Germany are shown in Table 2.1 for middle-class–working-class differences. Notice that the middle classes had an advantage of between 50 and 100 percent in these measures of reproductive fitness.

The higher socioeconomic classes also had an advantage in lower mortality. Some statistics for average ages of death in Switzerland, France and Germany in the eighteenth century have been calculated by Schultz (1991). Her results are shown in Table 2.2. The upper class consisted of aristocrats, professionals, senior civil servants, wealthy merchants and landowners; the middle class of teachers, shopkeepers, small merchants, and so forth; and the working class of skilled and unskilled artisans and servants. Note that the upper class had a longer life expectancy than the middle class, and these had a longer life expectancy than the working class. Since the average age of death was in the early to mid-twenties, many people died too young to have children, especially in the working class.

There is similar historical evidence from Japan for a positive association between socioeconomic status and fertility. Hayami (1980) divides the population into three classes and estimates that in the eighteenth and first half of the nineteenth centuries class 1 had 5.6 children, class 2 had 3.9 and class 3 had 3.7.

There were three major factors responsible for the higher mortality and lower fertility of the lower socioeconomic classes in historical times. These were, first, the use of infanticide and abortion, which are estimated to have been relatively widespread in preindustrial Europe (Coleman and Salt, 1992) and in Japan (Hayami, 1980), and which were more prevalent among the poorer classes who were less able to afford to rear children; second, the strong social controls preventing marriage for those without livelihoods and on sexual relations outside marriage, which reduces the fertility of the lowest classes; and third, the higher infant and child mortality of the lowest classes resulting from poor nutrition and greater exposure to disease resulting from overcrowded and unsanitary living conditions.

Table 2.1
Socioeconomic Differences in Fertility in Europe

Dates	Location	Middle Class	Working Class	Criterion	Reference
1560-99	England	4.1	3.0	Children born	Skipp, 1978
1620-24	England	4.4	2.1	Children born	Skipp, 1978
1625-49	England	4.0	3.4	Children born	Skipp, 1978
1650-74	England	3.8	3.4	Children born	Skipp, 1978
1547-1671	Saxony	3.4	1.6	Children marrying	Weiss, 1990
1500-1630	England	4.2	2.2	Children surviving	Pound, 1972

Table 2.2
Life Expectancy by Social Group

City	Period	Average age of death (in years)		
		Upper Class	Middle Class	Lower Class
Berlin	1710-1799	29.8	24.3	20.3
Geneva	17th century	35.9	24.7	18.3
Rouen	18th century	32.5	33.0	24.5
Neuruppin	1732-1830	33.2	28.6	28.9

Source: Schultz (1991).

The effect of poor nutrition on infant mortality has been shown in the twentieth century in a study of the effect of the Dutch famine of 1944–1945. The food intake in Western Holland was reduced to about 1,400 calories per day, approximately half that of the populations of economically developed nations in the twentieth century. The result was an approximately threefold increase in infant mortality as compared with 1939 (Lumey and van Poppel, 1994). Even a 10 to 15 percent weight loss in young women delays menarche and causes amenorrhea, which reduce fertility (Frisch, 1984). These factors together operated to keep the reproductive fitness of the upper middle classes substantially greater than that of the working classes.

9. SOCIAL MOBILITY IN PREINDUSTRIAL SOCIETIES

The inference that the greater reproductive fitness of the upper and middle classes, as compared with the lower classes, in historical times promoted natural selection for intelligence and character depends upon the assumption that there was a social-class gradient for these qualities. It is difficult to prove this conclusively but there is strong circumstantial evidence for believing that this was the case.

In any society where there is social mobility, those born with the qualities needed for upward social mobility tend to rise in the social hierarchy, while those born deficient in these qualities tend to fall. In all human societies the qualities for upward social mobility are intelligence and character. This was first recognized by Galton, who proposed that achievement depended on "ability" and character qualities which he described as "zeal" or "eagerness to work," and "capacity" or "the power of working" (1869, pp 78–79). Galton believed that anyone born with ability and zeal and capacity would rise in the social hierarchy.

In the twentieth century this formulation has become widely accepted in the social sciences. Michael Young (1958) in his book *The Rise of the Meritocracy* proposed the formula IQ + Motivation = Merit, and merit determined socioeconomic status. Later Jensen (1980, p. 241) rephrased the formula as Aptitude × Motivation × Opportunity = Achievement. There is a great deal of evidence in support of these formulas, which is reviewed in Chapters 11 and 13, and for the conclusion that both intelligence, work motivation and what Galton called character are substantially under genetic control, which is reviewed in Chapters 5 and 13.

It was argued by Galton and later by Ronald Fisher that in any society where there is social mobility the social classes will become to some degree genetically differentiated for intelligence, work motivation and other character qualities that determine social position. Fisher put this as follows in a letter written to E.B. Wilson in 1930 and quoted by Bennett (1983, p. 272):

> If desirable characters, intelligence, enterprise, understanding our fellow men, capacity to arouse their admiration or confidence, exert any advantage, then it follows that they will become correlated with social class. The more thoroughly we carry out the democratic program of giving equal opportunity to talent wherever it is found, the more thoroughly we ensure that genetic class differences in eugenic value shall be built up.

The argument of Galton and Fisher that in any society where there is social mobility the social classes must become genetically differentiated was restated in 1971 by Richard Herrnstein in his book *IQ in the Meritocracy.* Herrnstein put the argument in the form of a syllogism the terms of which were (1) if differences in mental ability are inherited; (2) if success in the socioeconomic hierarchy depends on these abilities; then (3) social-class differences will be based to some extent on genetic differences in these abilities. Herrnstein called this the specter of meritocracy. What he meant was that through social mobility the genes for high intelligence had come to be concentrated in the professional and middle classes, and the genes for low intelligence to be concentrated in the lower classes. Herrnstein's syllogism showed that logically this must be the case, granted the two premisses on which the syllogism was based. The argument applies equally strongly to the motivation and character contribution to social mobility.

The crucial question is, therefore, whether there was any social mobility in historical times. The classical work on this issue was written by Pitirim Sorokin, who surveyed a large number of historical societies and concluded that there has never been a society where there has been no mobility across social classes.

The nearest approach to a zero mobility society was the Indian caste system, but even that was occasionally penetrable. Apart from this "there has scarcely been any society whose strata were absolutely closed or in which vertical mobility was not present" (1927, p. 139). Sorokin showed in detail that in numerous societies throughout history gifted individuals rose in the social hierarchy and the ungifted fell. This was particularly the case during times of social change and disorder, which opened up opportunities for social advancement that were seized by talented individuals. For instance, in England in the sixteenth century the dissolution of the monasteries by Henry VIII provided opportunities to buy up the buildings and land, and in the seventeenth century the civil war provided further opportunities to buy up the estates of the royalists. In the eighteenth century there was a great expansion of the middle class, consisting of lawyers, army and navy officers, civil servants, bankers, merchants and manufacturers, described in detail by Stone and Stone (1986), and which continued to grow in the nineteenth century. Similarly in France, the revolution of 1789 was followed by the execution of many of the nobility and expropriation of their estates, opening up buying opportunities, and the creation of a new officer class and nobility by Napoleon.

The economist historian S.J. Payling (1992) concluded that there was significant social mobility in Europe from at least the fourteenth century. He described five routes by which talented young men from ordinary families moved up the social hierarchy. These were by becoming successful merchants; civil servants, such as tax collectors and administrators; lawyers; army and navy officers; and clergymen. Women were also socially mobile and moved up if they made advantageous marriages and down if their marriages were disadvantageous. Quantitative studies of the extent of social mobility have been made for the early nineteenth century. For instance, in Berlin in 1825, 20 percent of the men with working-class fathers entered middle-class occupations (Kaelble, 1985, p. 12).

The existence and extent of social mobility has also been shown in preindustrial China. From the second century B.C. the Chinese operated a system of competitive examinations for entry into the elite corps of mandarin administrators who governed the country and who enjoyed high social status. The examinations were difficult and success in them required intensive study for several years in universities. A description of the system has been provided by Bowman (1989). At least from the fourteenth century onward, there was a nationwide scholarship system which enabled poor but intelligent and well-motivated young men to attend the universities and work for entry to the mandarinate, and this provided a path of upward social mobility for able adolescents from all social classes. The family origins of 10,463 mandarins between the years 1371 and 1904 have been analyzed by Ho (1959). He found that 31 percent came from ordinary families, none of whose members had had degrees or offices for the three preceding generations. Sixty-three percent of mandarins came from middle-class and professional families, and the remaining 6 percent came from what were called "distinguished families" which had produced several mandarins.

The results show that most highly talented individuals in China came from the middle and professional classes, and this was an open elite which was maintained in each generation by recruiting into its ranks highly talented individuals from the lower classes.

Thus, for many centuries in Europe and China there has been significant social mobility in which talented individuals moved up the social hierarchy and the untalented moved down. These talented individuals in the higher social classes had greater reproductive fitness. Natural selection was working.

10. HIGH MORTALITY OF ILLEGITIMATES

A further way in which natural selection operated against low intelligence and weak character in historical times was by the exceptionally high mortality of illegitimate children. In spite of the strong social controls over extramarital sexual relations in Western nations until the mid–twentieth century, a small percentage of illegitimate births occurred. It is a reasonable inference that these were predominantly born to parents of low intelligence and weak character. There is direct evidence for this in studies carried out in the late twentieth century, which have shown that women who have illegitimate children are predominantly of below average intelligence, poorly educated, psychopathic and of low socioeconomic status. Herrnstein and Murray (1994) calculate that in the United States white, single welfare mothers have an average IQ of 92 as compared with 105 for those who are childless or are married with children. Women with less than a high school education are over twenty times more likely to have an illegitimate child than college-educated women (Rindfuss, Bumpass and John, 1980). Sexual promiscuity is one of the central features of the psychopathic personality, the extreme form of weak character, and likely in historical times to result in illegitimate children. It is easy to understand why single mothers tend to have low intelligence and weak character. They are less able to foresee, and they care less about, the adverse consequences of having an illegitimate child. These consequences were much more severe in historical times than in the late twentieth century, and it can be reasonably assumed that women who had illegitimate children, and the men who fathered these illegitimate children, were predominantly those of low intelligence and weak character, as they are today. Of course, not all of the parents of illegitimate children were of this kind. Intelligent women not infrequently became the mistresses of kings, aristocrats, cardinals and the wealthy. They were making a shrewd career move and their children were generally supported by their powerful fathers. But these were exceptions.

In historical times the state did not provide single mothers with welfare incomes and housing, as it has come to do in Western nations in the second half of the twentieth century. Single mothers generally abandoned their babies, and these generally died. In classical Rome, illegitimate and unwanted babies were put in the sewers, and anyone who wanted one had a large choice at their disposal (Dill, 1898). In later centuries in Europe, illegitimate and unwanted babies were abandoned in any convenient place, and up to the middle of the nineteenth century, it was not unusual to see dead babies in the streets and on rubbish dumps (Coleman and Salt, 1992). In the seventeenth and eighteenth centuries, the number of abandoned babies was so great in many European cities that orphanages were established to care for them. In Paris the Hospice des Enfants-Trouves was founded for this purpose in 1670 and took in several thousand babies a year, and in London the Thomas Coram Hospital was established for the same purpose in 1741. The death rate in these orphanages was high. In the Thomas Coram Hospital it is estimated that 71 percent of the children died by the age of fifteen, as compared with around 50 percent of the general population. The principal reason for the high mortality of the babies was that not enough wet nurses could be found for them, so they were inadequately nourished and particularly vulnerable to infectious disease.

Thus, in historical societies illegitimate children, born predominantly to parents with low intelligence and weak character, suffered high mortality. It was a cruel world, but it was a world in which the genes for low intelligence and weak character were constantly being expelled from the gene pool.

11. CONCLUSIONS

Natural selection operated in human societies up to the middle of the nineteenth century. Among the most primitive hunter-gatherer tribes, among pastoralists and agriculturalists, in the early nation-states and in Christendom, fertility and mortality were much higher than was needed for replacement; disease culled those with poor health and low social status and curtailed their fertility on an extensive scale.

There has always been social mobility in all human societies, through which those with intelligence and character—consisting particularly of the motivation and capacity for sustained work—have risen in the social hierarchy; those deficient in those qualities have fallen. Because intelligence and character are both significantly under genetic control, the reproductive fitness of the leaders and of the upper and middle classes ensured the operation of positive natural selection for these qualities over many centuries and millennia. It was not until the second half of the nineteenth century that natural selection broke down. How and why this occurred are the questions to which we turn next.

Chapter 3

The Breakdown of Natural Selection

1. The Decline in Mortality. 2. The Control of Infectious Diseases. 3. Dysgenic Effects of the Reduction of Infectious Diseases. 4. Residual Effects of Selection by Mortality. 5. Decline of Fertility. 6. Onset of Dysgenic Fertility. 7. Causes of Dysgenic Fertility. 8. Dysgenic Effects of Welfare. 9. Increase of Harmful Mutant Genes. 10. The End of Positive Natural Selection. 11. Conclusions.

We saw in the last chapter that natural selection operates throughout nature, and operated in human societies up to the nineteenth century to keep the populations genetically sound through the survival of the fittest. From the mid nineteenth century onward natural selection broke down in the economically developed nations, and this led to genetic deterioration with regard to health, intelligence and character. In this chapter we examine how the breakdown of natural selection came about.

1. THE DECLINE IN MORTALITY

Natural selection operates through high mortality and low fertility of the less fit. The first of these to weaken during the nineteenth century was high mortality. The reduction in mortality generally, and particularly in child mortality, began in the United States and Europe around the year 1800. This is illustrated by statistics of life expectation in England for the period 1750–1987 shown in Figure 3.1. Notice that during the years 1750–1800 life expectation averaged about 35 years. This does not mean that most people died in their thirties but rather that about half of those born died in infancy and childhood and the other half reached their fifties, sixties and seventies (Crow, 1966). From the year 1800 life expectation increased until it reached a little over 70 in the late twentieth

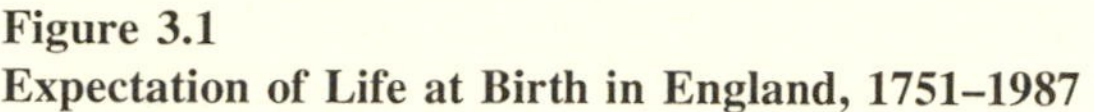

Figure 3.1
Expectation of Life at Birth in England, 1751–1987

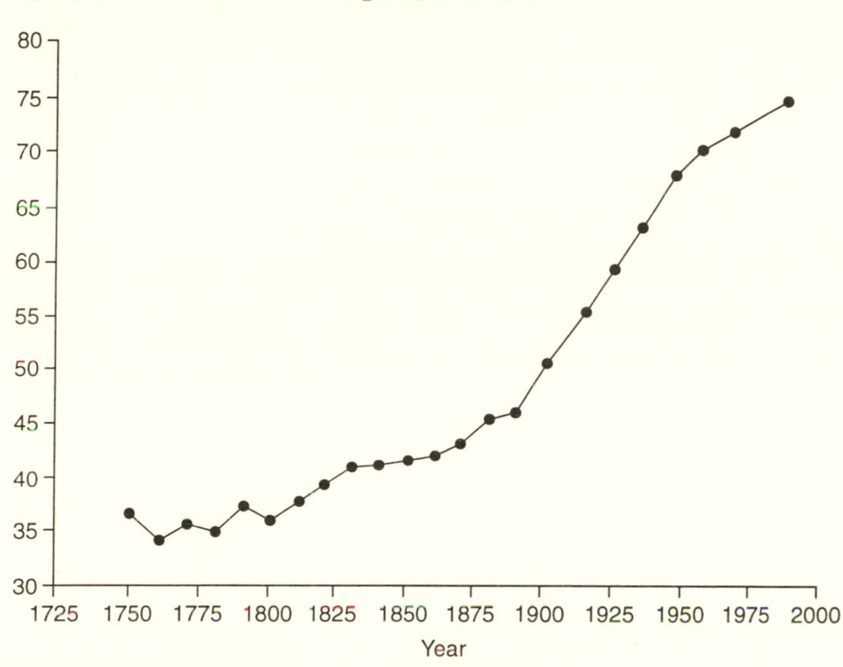

Source: Coleman and Salt (1992).

century. This increase in life expectancy occurred throughout the economically developed world.

Much of the greater expectation of life was achieved through the reduction of child mortality. For instance, in Sweden, where historical demographic records are particularly good, child mortality before the age of 15 was 43 percent in 1750; it fell to 34 percent in 1850 and to 5 percent in 1950 (Hed, 1987). By the 1990s infant and child mortality reached about 1 percent throughout the economically developed world. Virtually everyone survived to adulthood and was able to reproduce. Natural selection operating by the higher mortality of the less fit had disappeared.

2. THE CONTROL OF INFECTIOUS DISEASES

The principal cause of high mortality in historical times was infectious diseases, and it was the growing control of these that was the principal cause of the reduction of mortality. The most serious of the infectious diseases in Europe was bubonic plague which is carried by fleas on the black rat. The plague made its first recorded appearance in Istanbul in the year 542, when it killed about a quarter of the inhabitants. The worst outbreak was the Black Death of the early fourteenth century, when about two-fifths of the population of Europe died from the disease (McNeill, 1977). A number of further outbreaks occurred over the next four centuries. It has been estimated that about half the population of Seville died when the plague struck the city in 1654, and large number died in the outbreak in England in 1666, which was recorded by the diarist Samuel Pepys. Apart from the plague, the principal infectious diseases that caused high mortality were smallpox, malaria, tuberculosis, cholera, poliomyelitis, measles and leprosy. The early European colonists in the Americas introduced measles, and this had a devastating effect on the American Indians because so few of them had immunity to the disease.

From around 1800, mortality from infectious diseases gradually fell and by the late twentieth century had been reduced to negligible proportions. Smallpox has been eliminated entirely, and the other serious diseases have become effectively controlled. Three principal factors have been responsible for the reduction of mortality from infectious diseases. These were the development of immunization, improvements in nutrition and improvements in public health as a result of the supply of pure water and efficient sewage disposal.

The first advance in immunization was the development of vaccination against smallpox in the second half of the eighteenth century. Initially vaccination was used almost entirely by the better off, but in the first half of the nineteenth century it was made compulsory for all children throughout Europe and North America (Mercer, 1985), and this eliminated the disease in the economically developed world. More vaccinations were developed by Louis Pasteur in France in the 1860s and 1870s against diphtheria, tuberculosis, cholera, yellow fever and plague. The next major advance in combatting infectious diseases was the development of antibiotics and their widespread use from the middle of the twentieth century onward.

The second factor responsible for the reduction in mortality from around 1800 was the gradual improvement in nutrition. In England real living standards approximately doubled between 1800 and 1900, and doubled again between 1900 and 1990 (Coleman and Salt, 1992; Crafts and Mills, 1994; Himmelfarb, 1995), and similar improvements in living standards occurred throughout Western nations. One of the results of this was that people were able to eat better, so they became healthier, more resistant to disease and lived longer.

The third factor reducing mortality was the provision of pure drinking water and effective sewage disposal. These were progressively introduced throughout Western nations from the middle years of the nineteenth century. Before this, drinking water in towns was frequently contaminated by sewage and this was responsible for epidemics, particularly of cholera. By the middle of the twentieth century, mortality from infectious disease had become virtually eliminated in the economically developed world.

3. DYSGENIC EFFECTS OF THE REDUCTION OF INFECTIOUS DISEASES

The increasing control over infectious diseases had a dysgenic effect because those who previously died from these diseases were those who lacked immunities, those with poor general health, the less intelligent and those with weaker character. The relaxation of selection pressure against those who lacked specific immunities against particular diseases did not matter, because the diseases were controlled, but the increased survival of those with poor general health, the less intelligent and those with weak character had a more general dysgenic impact.

The reason for this was that those who had previously died from infectious diseases were disproportionately the poorer classes, who had lower nutritional status, and many of whom lived in unsanitary conditions in overcrowded and insalubrious towns and cities, where diseases were most virulent; these were less able to escape to the countryside when epidemics appeared. The lower classes tended to be less intelligent and have weaker character than the middle and upper classes as a result of centuries of social mobility, so their differentially high mortality from infectious diseases exerted selection pressure against low intelligence and weak character. As mortality from infectious disease declined in the nineteenth and twentieth centuries, this selection pressure weakened, and those with poor general health, low intelligence and weak character were the principal beneficiaries.

4. RESIDUAL EFFECTS OF SELECTION BY MORTALITY

Although natural selection, operating by the greater mortality of those with poor health, low intelligence and poor character, became greatly reduced during the nineteenth and twentieth centuries, it did not disappear entirely. In spite of medical advances, children born with genetic diseases and with poor general health continued to die more often than those without them, and there remained some greater premature mortality among the less intelligent and those with weak character.

Most of the evidence showing the continued operation of natural selection against low intelligence and poor character is derived from the indirect indices of educational level and social class. Illustrative figures of differences in infant mortality (deaths in the first year) by the educational level of their parents are shown for the United States for 1983–1985 in Table 3.1. These figures, based on approximately 10 million births of blacks and whites, are broken down by four educational levels and marital status. The table has three points of interest. First, for all four groups of married and unmarried blacks, and for married and unmarried whites, infant mortality is significantly greater among the least educated (those with less than twelve years of education) and declines steadily with higher levels of education. The explanation for this is not that education as such reduces infant mortality. Staying in high school and going to college, learning

Table 3.1
Infant Mortality Rates per 1,000 Live Births in the United States, 1983–1985

	Blacks		Whites	
Education	Married	Unmarried	Married	Unmarried
0-11 years	17.5	20.6	11.5	14.5
12 years	13.9	17.7	7.5	11.8
13-15 years	13.1	16.2	6.3	11.2
16+ years	11.1	16.5	5.6	8.3

Source: Bennett, Braveman, Egeter and Kiely (1994).

that Shakespeare wrote Hamlet, that the North won the American Civil War and so forth, does not in itself increase a mother's ability to keep her baby alive. Educational level is determined by intelligence and character, as shown in detail in Chapters 11 through 13, and it is having more intelligence and character that reduces child mortality.

The second point of interest is the greater incidence of infant mortality in the unmarried as compared with the married. This is probably not because having a husband in itself reduces infant mortality. The main reason for this difference is probably that the unmarried are less responsible, a component of their weak character, even when educational level is held constant. The third interesting feature of the table is the higher incidence of infant mortality among blacks. Probably the principal reason for this is the lower intelligence of blacks, even at the same educational levels. The data as a whole show that selection by infant mortality against the less intelligent and those with weak character was still present in the United States in the 1980s.

Somewhat similar data for socioeconomic-status differences in infant mortality and for adult mortality are shown for Britain in Table 3.2. The first column gives infant mortality rates for the five socioeconomic classes. Notice that infant mortality among the professional class is less than half that in the unskilled working class, reflecting the higher intelligence and stronger character of the professional class. The remaining four columns show standardized mortality ratios for males, in which the mortality for the total population is set at 100, for the years 1930, 1950, 1960 and 1970 for men aged 15 to 64. Notice that in all four years there is a gradient from low mortality in social class 1 increasing steadily to high mortality in social class 5. Notice also that the social-class differences were greater in 1960 and 1970 than they were in 1930 and 1950. A similar increase in mortality differentials between social classes has been found in the United States over the period 1960–1986 by Pappas, Queen, Hadden and

Table 3.2
Infant Mortality per 1,000 Live Births and Standardized Mortality Ratios for Men Aged 15–64 by Social Class in Britain

Social Class	Infant mortality 1981	Mortality, men aged 15-64 1930	1950	1960	1970
1. Professional	8.7	90	86	76	77
2. Managerial	9.6	94	92	81	81
3. Skilled Manual	10.8	97	101	100	104
4. Semiskilled	15.1	102	104	103	114
5. Unskilled	18.2	111	118	143	137

Source: Coleman and Salt (1992); Goldblatt (1990).

Fisher (1993). The reason for this is probably that a greater proportion of premature deaths are due to accidents and self-inflicted diseases caused by excessive smoking, drinking, eating and lack of exercise, for which poor intelligence and weak character are partly responsible.

Further data for socioeconomic-status differences in mortality are given for European countries for the 1970s by Leclerc (1989). He calculates the mortality of unskilled men in relation to 100 for all employed men. The ratios for 35 to 39-year-olds are Denmark (142), England (136), Finland (193), France (162) and Norway (130), showing considerable excess mortality among the unskilled.

The mortality gradient by educational level which has been noted in the United States has also been found in a number of countries. Kunst and Mackenbach (1994) have documented this for nine countries in the 1970s and 1980s. They calculated mortality for 35 to 44-year-old males in the least educated groups as a percentage of that in the most educated. The percentages of excess mortality among the least educated were 72 (Netherlands), 102 (Norway), 104 (England), 117 (Denmark), 120 (Sweden), 149 (Finland), 185 (Italy), 197 (France) and 262 (United States). The causes of these persistent social-class and educational-level differences in premature mortality have been debated by social and medical scientists. Kunst and Mackenbach (1994) conjecture that perhaps good health fosters high educational and occupational achievement, or alternatively that education reduces mortality. Both hypotheses seem implausible. Much of the explanation for the inverse association between socioeconomic status and mortality must be that the higher classes are more intelligent and more conscientious in their lifestyles. There is direct evidence for this in the case of intelligence. An unintelligent mother is more likely to lose a child through death because she is less alert to the dangers of accidents and infections and the importance of seeking medical help. Herrnstein and Murray (1994), in their analysis of the American Longitudinal Study of Youth sample, calculate that the mothers of children who died in their first year of life had an average IQ of 92—eight IQ points below the national average.

A study showing that low intelligence predicts premature death among adults has been carried out in Australia by O'Toole and Stankov (1992). They compared a sample of 523 national servicemen who died between the ages of 22 and 40 with a comparable sample of 1786 who survived. They found that those who died had average IQs about four points lower than those who survived, and poorer educational attainment, poorer occupational status and stronger psychopathic tendencies assessed by higher rates of criminal convictions. Both low intelligence and poor character contribute to premature death.

A significant factor in the greater premature mortality of the lower social classes is their greater accident-proneness. Data from Britain for accidents in the early 1980s show that deaths from accidents involving motor vehicles, fire, drowning and suffocation among one to nine-year-old children were about three times greater in social classes 4 and 5 as compared with 1 and 2 (Coleman and Salt, 1992, p. 359). Other major contributions to social-class differences in mortality are differences in cigarette smoking and alcoholism, both of which are substantially more prevalent among the lower social classes and significant causes of premature mortality (Leclerc, Lert and Fabien, 1990).

Thus natural selection by differential mortality has continued to operate against those with poor health, low intelligence and weak character in the late twentieth century, but it does so with greatly diminished force. Infants and children who die are easily replaced, and the magnitude of differential mortality in adolescence and early adulthood has become negligible. Most of the differential mortality by social class occurs in the second half of life, principally as a result of accidents, poor nutrition, smoking and excess alcohol consumption after the reproductive years are complete, so this has no selection effect. By the late twentieth century natural selection by mortality was still present, but it was of negligible proportions.

5. DECLINE OF FERTILITY

Natural selection operates through the twin processes of the fitter having low mortality and high fertility. We have seen that selection by differential mortality has largely disappeared since the mid-nineteenth century. We look now at the disappearance of the second arm of natural selection, differences in fertility.

The decline of both mortality and fertility, which occurred in the economically developed world during the nineteenth and early twentieth centuries, is known as the demographic transition. Before this, populations had high fertility and high mortality, and the net result was that the size of the populations remained approximately stable over time or increased slowly. By the time the demographic transition was more or less complete, toward the middle of the twentieth century

Table 3.3
Dates of the Beginning of the Decline of Marital Fertility by 10 Percent from Previous Maximum

Country	Date of Fertility Decline	Country	Date of Fertility Decline
France	1800	Denmark	1900
Belgium	1882	Norway	1904
Switzerland	1885	Austria	1908
Germany	1890	Finland	1910
Hungary	1890	Italy	1911
England	1892	Bulgaria	1912
Sweden	1892	Spain	1918
Scotland	1894	Ireland	1929
Netherlands	1897		

Source: Knodel and Van de Walle (1986).

in the economically developed nations, people had few children and only a few of them died. Once more, the size of the population remained approximately stable or increased only slowly, although the means by which this was achieved were different. High mortality and high fertility had been replaced by low mortality and low fertility.

Typically, although not invariably, the demographic transition took place in two stages. The first was a decline in mortality, especially in child mortality. This was followed a number of years later by a decline in fertility. The decline in fertility is generally considered by demographers to begin at the date in which marital fertility falls by 10 percent from its previous maximum. Using this criterion, Knodel and Van de Walle (1986) give the dates for the onset of fertility decline in seventeen European countries; these are shown in Table 3.3. Notice that, with the exception of France, these fertility declines nearly all took place in Europe in the short period between 1882 and 1900, and a similar decline occurred during these years in the United States. This demographic transition of the 10 percent fall in fertility from its previous high peak conceals a reduction of fertility among the higher socioeconomic classes which began several decades earlier in the nineteenth century. The much earlier decline of fertility in France, which occurred about the year 1800, has been a puzzle for historical demographers. It implies some Gallic know-how in the control of fertility which was not present in other Western countries, the precise nature of which need not detain us here.

Fertility declined gradually from its onset in the last two decades of the nineteenth century until the average two-child family emerged in the early post–World War II years. Later in the twentieth century fertility declined further and by the 1980s was generally below replacement throughout the economically developed world.

A possible ultimate scenario of the demographic transition would be the emergence of societies in which every individual had two children, all of whom survived to adulthood and had two children. In a society of this kind natural selection would no longer be present because there would be no differential mortality or fertility. Once natural selection had ceased, these populations would be genetically stable over generations. In the populations we are envisaging, the only disturbing force would be genetic mutations. The great majority of these would be for genetic disorders, low intelligence, mental retardation or poor character, because most genetic mutations are adverse. These adverse mutants would be transmitted to children and would gradually increase, causing a genetic deterioration of the populations. This would be a problem. However, the genetic deterioration that actually emerged as a result of the demographic transition was more serious than this because those with more intelligence and strong character began to have fewer children than those weaker in these qualities.

6. ONSET OF DYSGENIC FERTILITY

The reduction in fertility that took place in the economically developed nations from 1880 onward did not take place to the same extent among all individuals. It occurred most among the more intelligent, those with stronger character, in the professional and middle classes and among the better educated. Much of the body of this book is devoted to documenting this outcome in detail, but as a foretaste of what is to come we will look at completed fertility in relation to socioeconomic status in the United States for white married women born between 1840 and 1845 and between 1865 and 1870 and shown in Table 3.4. Notice that the professional and other middle classes had fewer children than the working classes, that there is a generally linear increase in fertility with declining socioeconomic status, and that this trend is more pronounced in the 1865–1870 birth cohort than in the 1840–1845 birth cohort. Dysgenic fertility, the tendency of those with low intelligence and poor character, indexed by socioeconomic status, had appeared in the United States in the birth cohort of 1840–1845 and become more serious in the later birth cohort. Natural selection against those with low intelligence and poor character had not merely declined. It had gone into reverse.

In order to provide an approximate measure of the magnitude of dysgenic fertility, the fertility of the lowest social class has been divided by that of the highest to provide a dysgenic fertility ratio. Notice that the dysgenic fertility

Table 3.4
Average Number of Children of Married White American Women Born 1840–1845 and 1865–1870 in Relation to Husband's Socioeconomic Status

Birth Cohort	Socioeconomic Status 1	2	3	4	5	Dysgenic ratio
1840-45	4.4	4.2	4.6	4.6	5.3	1.20
1865-70	3.0	3.5	4.2	4.5	4.9	1.63

Note: Socioeconomic status: 1 = professional, senior executive; 2 = lower white-collar occupations; 3 = skilled; 4 = semiskilled; 5 = unskilled.
Source: Kiser (1970).

ratio increased considerably over the two generations. Another way of looking at the data is to consider the greater fall in the fertility of women in the professional class, from 4.4 to 3.0, a reduction of 31 percent as compared with the fall in the fertility of the wives of unskilled workers, from 5.3 to 4.9, a reduction of 7 percent.

7. CAUSES OF DYSGENIC FERTILITY

Many books and papers have been written by demographers on why the populations of the economically developed nations began to reduce their fertility in the nineteenth century, and why the professional and middle classes reduced their fertility more than the working classes, and especially the unskilled working class. We shall not digress at length into the battles that have taken place over this problem, but we pause to note that one of the principal causes of the appearance of dysgenic fertility lay in the increase in knowledge and use of contraception.

The principal means by which information about contraception grew in the nineteenth century was through the publication of a series of books. The first influential book to provide detailed information on birth control in the English-speaking world was Richard Carlile's *Every Woman's Book*, published in London in 1826. It explained the use of the sponge—the condom made from sheep's gut or cloth—and withdrawal. In the United States, Robert Owen's *Moral Physiology* was published in New York in 1830 and provided the same information. In 1832 the American physician Charles Knowlton published what was to become the most widely read book on birth control in Britain and the United States in the middle years of the nineteenth century. This was *The Fruits of Philosophy* which, in spite of its somewhat opaque title, gave sound advice on the safe period, coitus interruptus, spermicidal chemicals, sponges, sheaths and a variety of pessaries and douches. In Britain this book sold about 100,000 copies in the middle decades of the century. It was reprinted in 1877 by Charles Bradlaugh and Annie Besant, for which they were prosecuted in a celebrated trial for the publication of obscene material. The trial generated a great deal of publicity and as a result sold 125,000 copies within three months. During the next three years a further 185,000 copies were sold, in addition to numerous pirated copies. Bradlaugh and Besant were acquitted, but in the same year Edward Truelove was tried and convicted for republishing Robert Owen's *Moral Physiology* and, as a predictable consequence, sales of the book soared. The year 1877 also saw the publication of Annie Besant's *The Law of Population*, an updated book on contraceptive methods which sold 175,000 copies over the next fourteen years, until the author converted to theosophy and withdrew the book from publication. A number of late-nineteenth-century observers dated the acceleration of the decline in the fertility of the reading classes from the widespread publicity received by these two books at the 1877 trials (Soloway, 1990).

It was in the 1870s that the modern condom was developed. This became possible as a result of Charles Goodyear's invention in the 1840s of the technology of the vulcanization of rubber by which lumps of raw rubber could be processed into thin sheets. In the early 1870s these became used for the mass production of condoms and in the 1880s for the production of the Dutch cap. These products quickly became marketed by mail order and through retail outlets. Initially they were used principally by the professional and middle classes because these were better informed about them, more intelligent, better long-term planners of their lives and could more easily afford them. This is the principal reason why fertility fell sharply among the professional and middle classes from the 1880s onward, fell less among the skilled working classes, and fell very little among the unskilled working classes, who were the most illiterate, the poorest, the least intelligent and had the weakest characters. As the twentieth century progressed, knowledge of these contraceptives and of subsequent contraceptive developments, such as the intrauterine device and the contraceptive pill, spread gradually to the working classes, with the result that their fertility also began to fall.

The uptake of contraception led initially to large social-class differentials in fertility in the closing decades of the nineteenth century and the early decades of the twentieth, followed by a lessening of this differential from the middle of the twentieth century onward. However, the differential has not been eliminated in any of the economically developed nations. The major reason for this lies in the inefficient use of contraception by the lower socioeconomic classes and the less educated. Even in the 1980s, when virtually everyone in the economically developed nations knew about contraception, large numbers of unplanned pregnancies and births continued to occur. For the United States, Kost and Forrest (1995) have analyzed the National Survey of Family Growth sample of women having babies in 1988 and found that 36 percent of them were unplanned. The incidence of unplanned births is strongly related to educational level. Among women with less than twelve years of education, 58 percent of births were unplanned, falling to 46 percent among those with twelve years of education, 39 percent among those with some college education and 27 percent among college graduates.

Unplanned pregnancies result from the inefficient use of contraception, and many studies have provided direct evidence that contraception is more efficiently used throughout the United States and Europe by the higher and more educated social classes than by the lower and less educated (Calhoun, 1991). For instance, in the United States, it has been reported by Forrest and Singh (1990) in a 1988 study that 72 percent of fifteen to nineteen-year-old women from middle-class families used contraception on the occasion of their first sexual intercourse as compared with 58 percent of those from poor families.

Once contraception became universally available it was inevitable that it would be used more efficiently by the more intelligent, those with stronger character expressed in greater self-control and the capacity to look ahead, and inefficiently by those with low intelligence and weak character, who would consequently have more children than they planned or desired. This is the main reason why dysgenic fertility is universal in the twentieth century except, as we shall see in Chapter 15, in a number of countries in sub-Saharan Africa where the use of contraception is negligible. Once contraception became widely available, dysgenic fertility became inevitable.

8. DYSGENIC EFFECTS OF WELFARE

Although the inefficient use of contraception by the lower social classes, the less educated, the least intelligent and those with weak character was the principal reason for the emergence and persistance of dysgenic fertility from the middle decades of the nineteenth century, it was not the sole reason. A contributing cause of dysgenic fertility has been the growth of the welfare state. Until the second half of the twentieth century there were strong disincentives for single women against having illegitimate children. Unmarried mothers almost invariably suffered loss of employment, loss of earnings and social stigma, and they were not given welfare payments and housing for themselves and their babies. Women understood this and the result was that the great majority of single women avoided becoming pregnant.

This began to change in the second half of the twentieth century as the governments of Western nations began to provide, or to increase substantially, welfare payments and housing for single women with children. The effect of this was to provide financial incentives for women with low intelligence, poor educational attainment and weak character to have babies, or at least to reduce the disincentives, and increasing numbers of them began to do so. The case that increasing welfare payments for single mothers had the effect of increasing the illegitimacy rate was first worked out for the United States by Charles Murray (1984) in his book *Losing Ground.* Murray showed that from the early 1960s the welfare benefits paid to poorly educated young women increased relative to the wages for unskilled work. There was also a reduction in the social stigma incurred by having babies out of wedlock. The combined impact of these factors was to increase the incentives and reduce the disincentives for having illegitimate children. The results have been that among white American women illegitimate births remained steady at 2 percent from 1920 to 1960, and then rose steadily to 22 percent in 1991; while among black women, illegitimate births were between 12 and 20 percent from 1920 to 1960, and then rose steadily to 67 percent in 1991 (Murray, 1984; Himmelfarb, 1995).

The increase in welfare benefits did not act as an incentive to have illegitimate children for women with high or average intelligence, good educational qualifications and strong character. The more attractive alternative for these women was to work in well-paid and fulfilling jobs and the last thing that appealed to them was to become single mothers. The increase in welfare benefits for single mothers provided a selective incentive to those with low intelligence, poor education and weak character. This is why it was among these women that the increase in single motherhood grew most strongly. This has been shown for the United States by Zill, Moore, Ward and Steif (1991) who estimate that over half of welfare single mothers are in the bottom 20 percent of the population for education attainment. Herrnstein and Murray (1994) in an analysis of another data set found that the average IQ of long-term welfare-dependent single women is 92.

It was not only in the United States that welfare benefits for single mothers were increased from the 1960s and 1970s onward and were followed by an increase in illegitimacy rates. The same thing happened in Australia, where welfare-dependent single mothers increased approximately fivefold in the 1980s (Tapper, 1990). Similar trends were widely present throughout Europe (Roll, 1992). In Britain, Murray's thesis has been confirmed by Patricia Morgan (1995). From the 1970s onward single mothers in Britain have received a basic income, free housing, remission of property tax, free health care and a variety of other benefits. The result has been an increase in illegitimacy, especially in the lowest social class, similar to that in the United States. Illegitimacy in Britain, which had stood at around 5 percent of births over the 400 years from 1550 to 1950, began to increase steadily from the 1970s and reached 32 percent in 1991 (Murray, 1994; Himmelfarb, 1995). As in the United States, the illegitimacy rate is highest in the lowest social class. Murray's (1994) analysis of the 1991 British census shows a correlation of .85 between the male unemployment rate and the female illegitimacy rate across local authorities, and that illegitimacy increased most over the years 1974 to 1991 among the lowest social class.

Murray's thesis that the increase of benefits for single mothers has increased the illegitimacy rate among the lowest social class has not been allowed to pass without criticism. The parallel increase in welfare payments to single mothers and of illegitimate births among the lowest social class does not prove cause and effect. In considering the plausibility of the Murray thesis, it is important to distinguish its strong and weak forms. The strong form states that significant numbers of young women with low intelligence and poor education in the lower socioeconomic classes are fully informed, rational calculators who have a precise knowledge of all welfare benefits and the alternatives of poorly paid employment, make a conscious decision that single motherhood is the better option and deliberately become pregnant. This is the form of the theory favored by Patricia Morgan in Britain who writes that "most unwed mothers conceive and deliver their babies deliberately, not accidentally" (1995, p. 75). The weak version of the theory is that possibly the majority of poorly educated single young women who become pregnant do so unintentionally, but at the same time they know from the experiences of their mothers, sisters and friends that having and rearing a baby is a viable and not an entirely unattractive option. This knowledge makes them less careful about using contraception than they would otherwise be, and once they are pregnant, less inclined to have the pregnancy terminated or put the baby up for adoption. This weaker version of the theory does not predict high correlations between the illegitimacy rate and the amount of welfare benefits available at a particular time and place which—although demonstrable in some studies, such as that of Plotnik (1992)—have generally been found to be weak (Jencks, 1992; Moffitt, 1992). This is the version of the theory that Murray himself prefers. The weaker form of the theory probably has some validity, so that a number of Western governments have themselves contributed to dysgenic fertility from the 1960s onward, by increasing the welfare incentives for women with low intelligence, weak character and poor education to have babies.

9. INCREASE OF HARMFUL MUTANT GENES

The genetic deterioration entailed by the reduction in mortality and increase in fertility of those with poor health, low intelligence and weak character is enhanced by the continued appearance in the population of new harmful mutant genes. A mutant is a new gene or, more commonly, a new allele, an alternative form of an existing gene which appears spontaneously and is therefore not inherited from the parents. The great majority of these mutant genes are harmful because the human body is so well-designed that any random alteration is much more likely to be adverse than beneficial. It has been estimated that a mutant appears in about one in every 100,000 genes, and that on average everyone has about one mutant (Neel, 1994).

When natural selection is working, individuals born with harmful mutants tend to have high mortality and low fertility, like the carriers of adverse genes inherited from their parents. The effect of this is that although harmful genes are constantly appearing in the population through mutation, they are equally constantly eliminated. With the breakdown of natural selection, the elimination of harmful mutants is much less effective and many of those born with them, who would formerly have died in infancy or childhood, are now surviving to adulthood and having children, some of whom inherit the gene and increase its prevalence in the population.

Not all harmful mutants are surviving in modern populations. There are still plenty of genetic diseases for which no treatment is available, so babies born with these mutants continue to die, and the gene is eliminated. But medical progress is increasingly able to treat genetic diseases, with the result that mutants for these diseases survive and spread in the population. Equally serious is the continued appearance of mutants for low intelligence and weak character. As noted in Chapter 2, in historical times the genes for these traits were eliminated from the population by high mortality and low fertility, but from the nineteenth century onward this process has largely ceased.

10. THE END OF POSITIVE NATURAL SELECTION

A further effect of the demographic transition and the breakdown of natural selection is that genetic improvement of the population by the appearance and spread of advantageous new mutants is no longer possible. Although most mutants are harmful, a few are beneficial. Under natural selection, an individual born with an advantageous new mutant tends to have lower than average mortality and higher than average fertility, so the mutant gradually spreads through the population. For instance, humans first evolved in central Africa and had black skin. Some of them migrated northward into Asia and Europe, and some of these migrants had mutants for white skin. These mutants were advantageous in colder latitudes because white skin allows the absorption of vitamin D from sunlight, whereas black skin is more advantageous in the tropics because it shields out strong ultraviolet radiation. Individuals in Eurasia with mutants for white skin tended to have lower mortality and higher fertility, so the new genes spread in the populations and ultimately replaced the genes for black skin.

Genetic improvements of this kind are no longer possible. Suppose that an advantageous new mutant for high intelligence appeared. It would be inherited by half of the individual's children. In contemporary societies most people have two children, so the advantageous new dominant would be inherited by an average of one child in each succeeding generation. With the end of natural selection there is no longer any mechanism for spreading the new gene in the population. When the geneticist Theodosius Dobzhansky delivered a series of lectures at Yale in 1962, he published them as *Mankind Evolving*, but in choosing this title he missed the whole point of the breakdown of natural selection, which is that mankind has ceased to evolve.

11. CONCLUSIONS

The function of natural selection of keeping populations genetically sound by high mortality and low fertility of individuals carrying undesirable genes has broken down in the economically developed nations in the nineteenth and twentieth centuries. The high mortality arm of natural selection broke down first, from around 1800, largely as a result of improvements in the control of infectious diseases, in public health and through better nutrition. The low fertility arm of natural selection broke down later, from around 1850, and went into reverse. From this time onward the less intelligent, those with weak character, the less educated and the lower social classes had high fertility, ushering in a period of dysgenic fertility which has persisted for more than a century. The principal cause of dysgenic fertility is more efficient use of contraception by the more intelligent, the better educated and those with stronger character. Dysgenic fertility has been exacerbated by the increase in welfare payments—providing incentives for single women lacking these characteristics to have babies—which many Western governments increased from the 1960s onward. The genetic deterioration brought about by these developments has been further augmented by the continued appearance of harmful mutant genes which were previously eliminated from the population by the high mortality and low fertility of those who carried them.

Natural selection can be thought of as a gardener who removes the weeds from the garden and occasionally adds a desirable new plant. The gardener keeps the garden in good order and every now and again improves it. When the gardener stops working the weeds proliferate unchecked and the garden deteriorates. During the course of the nineteenth century the gardener ceased to work in the Western nations. " 'Tis an unweeded garden that grows to seed." Hamlet said it all.

Chapter 4

The Genetic Deterioration of Health

1. Reduction of Immunities against Infectious Diseases. 2. Selection Effects of Sexually Transmitted Diseases. 3. Development of Surgical Treatments. 4. Further Medical Advances. 5. Treatment of Phenylketonuria. 6. Effect of Mutant Genes. 7. Increase of Genetic Diseases. 8. Conclusions.

The eugenicists believed that the breakdown of natural selection was leading to the genetic deterioration of the populations of Western nations in respect of health, intelligence and character. This thesis has been sketched in outline and we are now ready to look at it more closely. The argument is most straightforward as it applies to health, and we shall consider it first. The basic point is that medical progress in the twentieth century has made it possible to treat many of those with genetic diseases, disorders and disabilities, and enabled them to survive to adulthood and have children. The result is that the genes for these diseases and disorders are no longer being so effectively eliminated, and their incidence in the population is increasing.

1. REDUCTION OF IMMUNITIES AGAINST INFECTIOUS DISEASES

We noted in the last chapter that from around the year 1800, mortality from infectious diseases began to fall as a result of the development of immunization, improvements in public health and better nutrition. From the point of health, the effect of this was that those lacking immunities to these diseases, many of whom would previously have died, survived in greater number. In one sense this was a genetic deterioration because the populations have become less resistant to these diseases. However, this is not a serious problem now that these diseases no longer pose any appreciable threat to Western populations.

There is no advantage to having a genetic immunity to smallpox, which no longer exists, or to most of the common infectious diseases which are controlled, largely through the immunization of babies or—if they are contracted—by antibiotics. Occasionally there are outbreaks of these infectious diseases, such as tuberculosis among the poor, but they are not a major problem in the economically developed world. The loss of immunities against the common infectious diseases is not a genetic change in Western populations that need cause concern.

2. SELECTION EFFECTS OF SEXUALLY TRANSMITTED DISEASES

Although most of the serious infectious diseases are effectively controlled in the economically developed nations, a number of the sexually transmitted diseases are still relatively common and are having a selection impact. The most important of these are chlamydia, gonorrhea, syphilis and AIDS (Acquired Immune Deficiency Syndrome). These diseases carry a significant risk of infertility and, in the case of AIDS, early death, and they consequently exert selection pressure against those who contract them and lack immunities.

These diseases also continue to exert selection pressure against those with low intelligence and weak character, because these people are at greater risk of contracting them and, if this occurs, not getting treatment. The diseases are contracted primarily through promiscuous sexual behavior, by not using condoms and, in the case of AIDS, by intravenous drug injections. These behaviors are more common among the less intelligent and those with weak character who do not understand, or do not care about, their future health and fertility. A useful study illustrating these points has been carried out in the Netherlands by Prins (1994). Patients at a Sexually Transmitted Disease Clinic who had had at least five sexual partners during the six months preceding their first visit were monitored over a two-year period. In spite of receiving medical advice at the Clinic, 49 percent of the women and 29 percent of the men contracted one of the sexually transmitted diseases. The most common was chlamydia, which was contracted by 32 percent of the women and 20 percent of the men. Gonorrhea and syphilis were contracted by 9 percent and 3 percent of the men, and 3 percent and 1 percent of the women, respectively. As would be expected, women who engaged in unprotected intercourse were 2.2 times more likely to have acquired one of the diseases.

It might be supposed that when individuals contract one of the sexually transmitted diseases, visit a clinic for treatment and are informed about the risks to their health and fertility, they would use condoms in subsequent sexual encounters. Yet, the Dutch study shows that this is not the case. The same conclusion has to be drawn from an American study by O'Donnell, Doval, Duran and O'Donnell (1995). They report an investigation of 691 black and Hispanic patients attending a New York STD clinic and given coupons for condoms for future use. Only 22 percent used the coupons, implying that most of those who contract sexually transmitted diseases are not concerned about the likelihood of further infections.

It is difficult to find direct evidence that those who contract sexually transmitted diseases are disproportionately the less intelligent and of weak character. The reason for this is that intelligence and character have become taboo concepts among the researchers in epidemiology, sociology and demography. Nevertheless, there is direct evidence that risky sexual behavior occurs predominantly among those with poor school achievement and among the sexually promiscuous. For instance, Luster and Small (1994) in a study of 2,567 13 to 19-year-olds in the American Midwest found that those who did not use condoms and had multiple partners had poor school grades and greater excess alcohol consumption than those who were more careful. Poor school grades and excess alcohol consumption are indirect measures for low intelligence and weak character, as will be seen in detail in Chapters 9 and 12. Confirmation of this American evidence that the less educated are less likely to use condoms comes from a study in France. Moatti, Bajos, Durbec, Menard and Serrand (1991) report the results of a survey of 1,008 individuals carried out in 1987. Among college graduates, 52 percent used condoms, falling to 47 percent among high-school graduates, and 43 percent among those with basic education.

There is also a strong relationship between the incidence of sexually transmitted diseases and socioeconomic status. For instance, a survey of the incidence of gonorrhea in Seattle carried out in 1987 and 1988 found incidence rates of 183 per 100,000 in the upper socioeconomic group, rising to 224, 421 and 1,147 in upper-middle, lower-middle and low socioeconomic-status groups, respectively (Rice, Roberts, Handsfield and Holmes, 1991). Socioeconomic status is correlated with intelligence and character, as we shall see in detail in Chapters 11 and 12, so the study provides indirect evidence for a high incidence of gonorrhea, and risk of infertility, among those with low intelligence and weak character.

The most serious of the sexually transmitted diseases is AIDS, caused by the Human Immunodeficiency Virus (HIV). Apart from the homosexual community, AIDS is most prevalent among the lowest socioeconomic classes (Bell, 1989; Shaw and Paleo, 1986), criminals (Zimmerman, Martin and Vlahov, 1991), drug-injectors (Chamberland and Donders, 1987; Rogers, 1987) and the sexually promiscuous (Mason et al., 1987). All of these categories are disproportionately of low intelligence and weak character, and it can be inferred that these have relatively high death rates from AIDS.

Because several of the sexually transmitted diseases, notably chlamydia, gonorrhea and syphilis, carry an appreciable risk of infertility; because AIDS leads to early death, and because those with low intelligence and weak character are more likely to contract these diseases and become infertile or die, natural selection is still having some impact against those with low intelligence and weak character. As modern medicine becomes more effective in the treatment of these diseases, natural selection against those with low intelligence and weak character will weaken further.

3. DEVELOPMENT OF SURGICAL TREATMENTS

From the early years of the twentieth century surgical treatments have been developed for numerous genetic diseases and disorders. In 1912 a surgical treatment was developed for congenital pyloric stenosis, a genetic defect of the stomach occurring in about one in 4,000 babies. The condition consists of a narrowing of the outlet which prevents food from being ingested, and the treatment consists of the surgical widening of the outlet. Before the treatment was developed, babies born with the defect died. The treatment enabled babies to live and transmit the defective gene to their descendants.

Another example of a genetic disorder that has become treatable by surgery is the eye cancer retinoblastoma. Up to the middle of the twentieth century, babies born with the cancer invariably died in childhood. In the second half of the century the condition has been treated by removal of the eyes. This enables the affected individual to survive and have children. The gene is dominant, so half of these children inherit it, have the defect and undergo surgery for the removal of their eyes. Another genetic disorder that has become treatable by surgery is defects of the heart.

A further development of surgical interventions has been organ transplants, in which a defective organ or tissue is replaced with a healthy substitute, usually taken from a dead person. The first successful transplant was corneal grafting and was carried out early in the twentieth century. In the second half of the century transplants were made of the kidneys, liver, pancreas, heart and lungs. All these surgical treatments have had the effect of prolonging the lives of those who formerly would have died, enabling them to have children and preserve the defective gene in the population.

4. FURTHER MEDICAL ADVANCES

A variety of other medical treatments for genetic diseases and disorders have been developed during the course of the twentieth century. In 1922 the discovery of insulin made it possible to treat insulin-dependent diabetes. This disorder is present in about 1.4 persons per 1,000 in Western populations and is caused by insufficient production of insulin. From 1922 onward it became treatable by taking insulin supplements.

The most common genetic disease of Caucasian populations in Europe and North America is cystic fibrosis. It is a recessive gene disorder and about 5 percent of Caucasians are carriers. This does them no harm because the recessive allele is negated by its healthy dominant. However, if two carriers have children, an average of one in four inherits the double recessive and has the disease. This occurs in about one baby per 2,000 births. The principal symptoms of cystic fibrosis are chronic lung infections and disorders of the digestive tract, liver and pancreas. Until the middle of the twentieth century, babies born with the disease invariably died in infancy or childhood, but from that time onward increasing number have been treated with a range of antibiotics and, more recently, by lung transplants. The result of these treatments has been that by the last decade of the twentieth century, about two thirds of those with cystic fibrosis survived into their twenties. They can and do have children, of whom an average of half inherit the gene.

Another recessive gene disorder for which medical progress has extended the life span of sufferers—but not to full term—is thalassemia, a form of anemia common in Southeast Europe. Formerly, sufferers died in childhood, but today repeated blood transfusions frequently prolong their lives into early adulthood. A further instance of a relatively common genetic disease that became treatable in the middle of the twentieth century is hemophilia. This is a blood-clotting disorder inherited through an x-linked recessive, by which sons inherit the disease from their mothers, who are carriers of the gene but do not have the disease. Because hemophiliacs lack a blood-clotting mechanism, virtually all of them bled to death at some time during their childhood. In the middle decades of the twentieth century, the disease became treatable when the bleeding could be stopped by an injection of a clotting agent known as Factor VIII. If necessary, blood transfusions could be given. The result has been that most hemophiliacs now live a normal life span and have children. If they mate with an unaffected female, their daughters inherit the gene as a recessive and become carriers, keeping the gene in the population.

5. TREATMENT OF PHENYLKETONURIA

One of the most successful medical treatments for serious genetic disorders to be developed in the twentieth century has been for phenylketonuria (PKU). This disorder consists of a defect in the process by which an enzyme converts phenylalanine (an amino acid) into tyrosine. The result is that phenylalanine builds up in the body. Normally this causes severe mental retardation, although sometimes the retardation is mild. Until the middle decades of the twentieth century, those who inherited phenylketonuria generally died young or, if they survived, did not have children. Either way, most of them did not reproduce, so the gene did not spread.

The genetics of phenylketonuria were worked out in the 1930s by the Norwegian physician I.A. Folling, who showed that the condition is caused by a recessed gene. If two carriers of the gene have children, the two recessives come together in an average of a quarter of the offspring, in accordance with the laws of Mendelian inheritance, and the child has phenylketonuria. This occurs in about one baby in 16,000 in Caucasian populations. Folling also showed that because PKU is caused by a build-up of phenylalanine, it can be treated by eliminating phenylalanine from the diet. Unfortunately phenylalanine is present in milk, meat and most protein foods. Hence, the affected babies and children have to be put on a very low-protein, mainly vegetarian, diet to prevent the onset of mental retardation.

In the early 1960s a simple blood test was developed by Guthrie which could be given to babies to test for the presence of PKU. Within the next few years the Guthrie test came to be given routinely to babies in most American states, in Britain and in Continental Europe. If PKU is diagnosed, mothers are advised to put their children on a phenylalanine-free diet and if this is done, the affected individuals survive with near normal intelligence and can have children.

It is sometimes believed that these medical advances have provided a complete cure for phenylketonuria so that the disease is no longer a problem. This is far from the case. There are several problems. First, parents find it difficult to exercise complete control over the children's diet, and the affected children frequently obtain some phenylalanine which impairs their intelligence. This is probably the main reason why the intelligence of those with the disease is typically somewhat below normal. For instance, a study of 51 affected children who were taking the protein-reduced diet found that their average IQ was seven IQ points lower than that of their unaffected siblings at 5 to 7 years of age (Fishler, Azen, Friedman and Koch, 1989). A second problem is that treated females pass the condition on to their children during pregnancy, and this usually produces retarded children (Schultz, 1983). The reason for this is that the metabolic disorders associated with PKU are not completely corrected by a phenylalanine-free diet.

Thus, the treatment of phenylketonuria is not completely effective. Some degree of impairment remains, although the treatment is sufficiently effective to allow the sufferers to survive and have children. If the partner does not carry the defective gene, all the children inherit the gene as a single recessive. If the partner is a carrier, half the children inherit the single recessive, and the other half inherit the double recessive and have the disease. If the partner has the disease, all the children inherit the double recessive and have the disease. Whatever the genetic status of the partner, the defective gene remains in the population.

6. EFFECT OF MUTANT GENES

Medical treatments for genetic diseases do not in themselves lead to an increase in the genes for these diseases in the population. These treatments prevent the genes from being eliminated, or reduce the extent to which they are eliminated; but if everyone with the defective gene has two children who survive to adulthood, the genetic structure of the population remains stable. What causes deterioration is the accumulation of new mutants for the disease.

It is estimated that on average most people have one mutant gene. Many of these do not do much harm and those who have them are unaware of their existence. Others are for only marginally inconvenient characteristics, such as

Table 4.1
Estimated Rates of Increase of Five Genetic Diseases over 30 Years

Disorder	Life expectation	Percentage increase in 30 years
Huntington's chorea	55	0
Duchenne muscular dystrophy	18	0
Hemophilia	70	26
Cystic fibrosis	22	120
Phenylketonuria	70	300

Source: Modell and Kuliev (1989).

having webbed feet or only three fingers. Others are more serious but not lethal, such as dwarfism which is caused by a dominant gene that appears relatively frequently by mutation. More serious still are the mutants for genetic diseases like cystic fibrosis, sickle cell anemia and the like, which severely impair the quality of life.

Estimates have been made by Neel (1994) of the rate of appearance of mutants for twenty of the most common genetic diseases. He concludes that the incidence is about four per 100,000 births for each disease. Thus about one live birth per 1,000 has a mutant for one of these diseases. There are in total around 4,000 identified genetic diseases for which mutants can occur (McKusick, 1992), so the number of newborns with one of these disorders is substantially greater and is somewhere on the order of 1 percent of births.

It was suggested by Muller (1950) that the rate of mutation is increasing in industrialized societies as a result of greater exposure to atomic radiation and industrial chemicals. This is probably so, but the magnitude of the problem is unknown. What appears certain is that the incidence of genetic diseases and disorders will gradually increase as a result of mutations and the treatment of those who formerly would have died. With further medical progress, this problem will inevitably become serious.

7. INCREASE OF GENETIC DISEASES

Although it is difficult to estimate the increase in the incidence of all the genetic diseases resulting from progress in medical treatment and the appearance of mutants, estimates for the increase of five of the more common have been made by the medical geneticists Bernadette Modell and A.M. Kuliev (1989). Their figures for the life expectation and percentage increase over one generation for Huntington's chorea, Duchenne muscular dystrophy, hemophilia, cystic fibrosis and phenylketonuria are shown in Table 4.1. The estimates are based on the life expectation of the sufferers, the rate of appearance of new mutations for the disorders, and whether the disorder is caused by a dominant, recessive or x-linked gene. Huntington's chorea is a disease that typically does not appear until early middle age and is characterized by physical and mental deterioration and death about ten years after its first onset. It is inherited through a dominant gene, so that an average of half the children of those with the gene inherit it. It is not estimated to increase at all because the mutation rate is relatively low, and those who have the disease are estimated to have fewer than an average of two children. Thus, the increase in the disease caused by new mutation is roughly counterbalanced by the below average fertility of those with the disorder.

Duchenne muscular dystrophy is caused by an x-linked recessive inherited by males from their mothers. It is not expected to increase because the sufferers are generally confined to a wheelchair from the age of 10 to 12, die in late adolescence and are very unlikely to have children. The gene is still being eliminated by natural selection but constantly reappears through new mutations.

Hemophilia is projected to increase by 26 percent because sufferers are able to live a normal life span and have children. Cystic fibrosis is projected to increase by 120 percent because of the prolonged lives of those affected, and because the gene appears relatively frequently by mutation. Finally, phenylketonuria is projected to triple every generation because the gene appears relatively frequently by mutation, and nearly all of those affected are now enabled to live a normal life span and have children.

A useful study illustrating the success of modern medicine in preserving the lives of those with genetic diseases and disorders who formerly would have died has been made in Denmark. Dupont (1989) has analyzed census returns for the prevalence of severe mental retardation in 1888 and 1979, and her results are shown in Figure 4.1. Notice that the prevalence rates are approximately twice as great in 1979 as they were in 1888, except between the ages of 5 and 12, when they were about the same. Many of the severely mentally retarded are infertile, so their increased survival into adulthood is not a major genetic problem—although some of them can and do have children.

8. CONCLUSIONS

Until the twentieth century, those born with serious genetic diseases and disorders generally died in childhood, so the genes for these conditions were gradually eliminated from the population. During the twentieth century, medical treatments have been developed for increasing numbers of genetic diseases and disorders. These consist of surgery, antibiotics and various treatments for specific disabilities. The effect of these treatments has been to enable many of those with these diseases to survive into adulthood and have children, to whom they transmit their defective genes. The deterioration of the gene pool is caused by the constant appearance of mutant genes for these diseases and disorders. The result is that the incidence of some of these diseases is doubling or even tripling in every generation.

Figure 4.1
Age-Specific Prevalence for the Mentally Retarded, 1888 and 1979 (Percent of the Population)

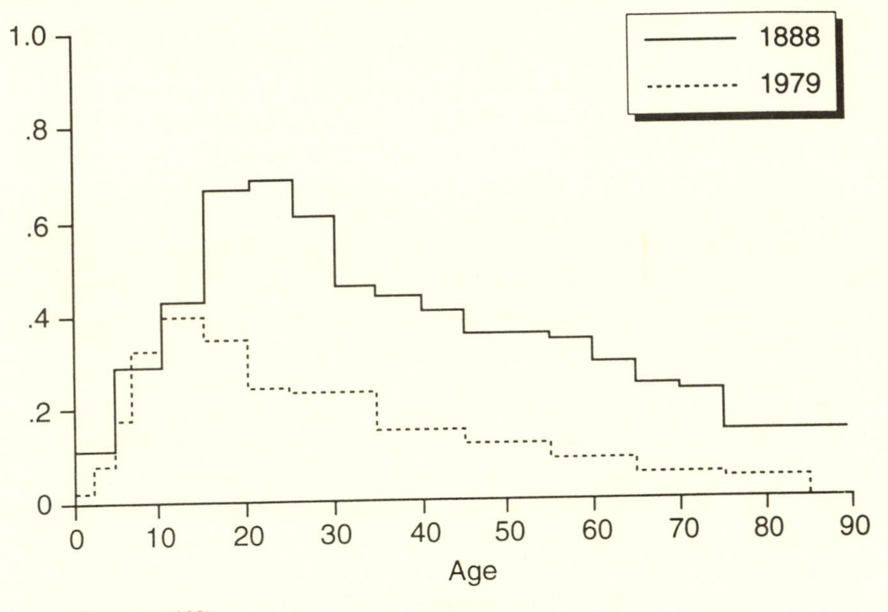

Source: Dupont (1989).

The problem of the increase of genetic diseases and disorders is likely to become worse in the future. This is partly because higher levels of radiation and the growing use of chemicals in industry and agriculture are probably increasing the rate at which harmful genes are appearing by genetic mutations. In addition, the pace of medical advance is rapid. Virtually every year medical treatments are improved and new methods developed. Medical progress has meant, and will continue to mean, genetic deterioration.

Chapter 5

Intelligence and Fertility: Sibling Studies

1. Inverse Relationship between Intelligence and Number of Siblings. 2. Lentz Calculates the Rate of Intelligence Decline. 3. Estimates of the Rate of Decline of Intelligence in Britain in the 1930s and 1940s. 4. Who Are the Childless? 5. Does Large Family Size Adversely Affect IQ? 6. The Decline of Intelligence in Britain, 1955–1980. 7. Estimating the Decline of Genotypic Intelligence. 8. Conclusions.

In the first decade of the century, the eugenicists suspected that intelligent people were having fewer children than the unintelligent, and therefore that fertility for intelligence was dysgenic. With the development of intelligence tests, it became possible to examine this issue and quantify its magnitude. As people thought about the problem, two methods of tackling it emerged. The first of these consisted of an examination of the relationship between individuals' IQs and their number of siblings, and we shall look at this evidence in the present chapter. The second was the relationship between people's IQs and their number of children, and we shall consider this evidence in the next two chapters.

1. INVERSE RELATIONSHIP BETWEEN INTELLIGENCE AND NUMBER OF SIBLINGS

In the 1920s studies began to appear concerning the relationship between IQs and number of siblings, first in the United States and Britain and subsequently in Continental Europe and elsewhere in the economically developed world. All of these studies, mainly but not invariably carried out on schoolchildren, found that the correlation between IQ and sibling size was negative: the higher a child's IQ, the fewer the number of his or her siblings. The results of a number of the leading studies are summarized in Table 5.1. Notice that all the correlations are negative; they range between −.19 and −.34 and average around −.26.

Those who collected these data from the 1920s onward believed that in general children's IQs are similar to those of their parents. Hence, they argued that the fact that children with low IQs typically had a lot of siblings must mean that parents with low IQs have a lot of children. They argued that if this were so, the intelligence of the population must be declining.

2. LENTZ CALCULATES THE RATE OF INTELLIGENCE DECLINE

It was a brilliant but now almost entirely forgotten American psychologist named Theodore Lentz (1927) who first devised a method for estimating the magnitude of the decline in intelligence from the negative association between children's IQs and their sibship size. The ingenuity of the method was that it achieved the apparently impossible feat of calculating the average IQ of the parental and the child generations from data derived solely from the IQs of children and number of siblings. The calculation of the average IQ of the parental and the child generations showed that the parental generation had a higher mean IQ than the child, and the difference between the two represented the decline in intelligence over the course of a generation. Lentz's reasoning was that each child has, on average, the same IQ as its parents and its siblings. He argued that when data on the IQs of schoolchildren and their number of siblings are collected, the parents of one child are less likely to appear in the sample than parents of several children. For instance, parents with six children are six times more likely to appear in such a sample of children than parents of one child. A statistical adjustment needed to be made for this undersampling of parents with few children. This was done by weighting the parental IQs (assumed to be the same as those of their children) by their number of children. This adjustment raises the average IQ of the parental generation. Using this technique, Lentz calculated from his data that the mean IQ of the parental generation was four points higher than that of the children's, indicating a four IQ point decline in one generation.

3. ESTIMATES OF THE RATE OF DECLINE OF INTELLIGENCE IN BRITAIN IN THE 1930s AND 1940s

Lentz's method of calculating the generational decline in intelligence from the negative association of IQ with number of siblings was taken up in Britain in the 1930s and 1940s. The first British study on this question was carried out in the mid-1930s by Cattell (1937) on 3,734 11-year-olds. The generational decline was calculated at 3.2 IQ points. However, this was an overestimate because Cattell's test had a standard deviation of 22 rather than the customary 15 or 16 (at this period the convention of setting the standard deviation of the

Table 5.1
Negative Correlations between IQ and Number of Siblings

Country	Number	Age	IQ x N Siblings	Reference
United States	629	12-14	-.33	Chapman & Wiggins, 1925
" "	4,330	6-20	-.30	Lentz, 1927
" "	554	5-23	-.19	Thurstone & Jenkins, 1931
" "	1,140	14	-.22	Burks & Jones, 1935
" "	156	13-18	-.31	Damrin, 1949
" "	979	Adults	-.26	Bajema, 1963
" "	12,120	Adults	-.29	Van Court & Bean, 1985
Britain	393	10	-.25	Bradford, 1925
"	1,084	10-11	-.22	Sutherland & Thomson, 1926
"	581	11-13	-.23	Sutherland, 1930
"	3,305	9-13	-.22	Roberts, Norman & Griffiths, 1938
"	10,159	9-12	-.23	Moshinsky, 1939
"	70,200	11	-.28	Thomson, 1949
"	9,183	18	-.34	Vernon, 1951
"	7,416	11	-.32	Nisbet, 1958
Greece	295	7-12	-.40	Papavassiliou, 1954
New Zealand	849	11-12	-.16	Giles-Bernardelli, 1950

intelligence scale at 15 or 16 was not firmly established). When Cattell's result is adjusted for the large standard deviation of the test, the rate of decline he obtained is reduced to 2.2 IQ points per generation.

In the late 1940s this problem was addressed again in Britain by Cyril Burt (1946) and Godfrey Thomson (1946), the Scottish psychologist who was largely responsible for the two massive surveys of the intelligence of all Scottish 10 and 11-year-olds in 1932 and 1947. They worked out a modified version of Lentz's method of calculating the generational decline. Like Lentz, they started with a sample of children and the negative correlation between their IQs and their number of siblings. They also followed Lentz in the assumption that the average IQ of the parents and of the siblings was the same as that of the child whose IQ was tested. Next, it was argued that most of these parents had other children outside the sample. The IQs of these would be, on average, the same as those of their siblings who were tested. The mean IQs of all these untested siblings could be calculated by weighting each child's IQ by the number of siblings and the number of families of this family size. The difference between the average parental IQ and the average IQ of all their children represented the generational decline.

Burt (1946) used this method on data he had collected in London and calculated the generational decline at 1.9 IQ points. Thomson (1947) applied the same method to calculate the generational decline for the sample of 11-year-olds whose IQ and sibling size data he had collected in 1925 and reported in Sutherland and Thomson (1926). The mean IQ of the children was 101.04, which he took as a measure of the average IQ of their parents. The mean IQ of all the children, including the untested siblings, was 98.98, and the difference of 2.06 between the two figures represented the generational decline. Although this method of estimating the generational decline in intelligence differed from that originally worked out by Lentz, Thomson demonstrated that the two methods yielded closely similar results.

A further study by Thomson used the same method to calculate the generational decline of intelligence in the city of Bath from a large scale data set collected by Roberts, Norman and Griffiths (1938). His calculation concluded that the Bath data showed a decline of 2.04 IQ points per generation. Thus, the four major studies of this question in Britain carried out in the 1930s and 1940s by Cattell, Burt and Thomson showed close agreement in estimating the rates of decline of the national intelligence at 2.20, 1.90, 2.06 and 2.04 IQ points per generation. These can be averaged to 2.0 IQ points to represent the intergenerational rate of decline in Britain in the first half of the twentieth century. I believe that this method was valid and the conclusion about right. It was, however, subjected to two criticisms and these need to be discussed. We will consider them in the next two sections.

Table 5.2
Percentages of British Males and Females Childless at Age 32 by IQ Groups

	IQ		
	Low	Average	High
Males	24	24	28
Females	11	16	18

Source: Kiernan and Diamond (1982).

4. WHO ARE THE CHILDLESS?

As the second half of the century unfolded, doubts began to be voiced about the validity of this method for measuring the rate of decline of intelligence. Two objections were raised. The first was that because the data for making the calculations are derived from children, the sampling procedure does not include individuals in the parental generation who are childless. This problem was recognized by people like Cattell, Burt and Thomson, but they believed that the childless had above-average intelligence, and they cited data to support this view. On these grounds, they believed that the inclusion of the childless in the calculations would have increased the size of the decline.

The assumption that the childless were of above-average intelligence was challenged in the early 1960s in two influential papers, one by Higgins, Reed and Reed (1962) and the other by Bajema (1963). Both these studies appeared to show that, contrary to the earlier assumption, childlessness was most prevalent among those with low intelligence and that if these were included in the calculation, there was no decline in the intelligence of the population.

These two studies were, however, on small and unrepresentative samples, and subsequent studies on larger and more representative samples have indicated that the childless are predominantly the more intelligent and, in addition, the better educated. The first of these studies was published for Britain by Kiernan and Diamond (1982) in an analysis of the data from a national longitudinal cohort study of 13,687 infants born in 1946. The proportions childless at age 32 analyzed by low, average and high IQ categories are shown in Table 5.2. Notice that for both males and females the greatest proportions of childless are in the high IQ group and that the tendency of childlessness to increase among higher IQ groups is more pronounced for females than for males. This is consistent with a number of studies using a variety of different methodologies pointing to greater dysgenic fertility among females than among males.

Similar results have been published for the United States by Van Court (1985) based on a large representative national sample of around 12,000 adults. Intel-

Table 5.3
Comparison of Vocabulary Scores of Those with Children and Those without

Birth cohort		Without children Mean	With children Mean
1	1890 - 1894	5.3	5.2
2	1895 - 1899	5.6	5.5
3	1900 - 1904	6.7	5.4
4	1905 - 1909	5.7	5.4
5	1910 - 1914	6.2	6.2
6	1915 - 1919	6.7	6.1
7	1920 - 1924	6.1	6.1
8	1925 - 1929	6.5	6.2
9	1930 - 1934	6.3	6.2
10	1935 - 1939	6.4	6.1
11	1940 - 1944	6.5	6.3
12	1945 - 1949	7.0	6.1
13	1950 - 1954	6.3	5.4
14	1955 - 1959	5.7	4.7
15	1960 - 1964	5.2	4.5

Source: Van Court (1985).

ligence was measured by a vocabulary test and the sample divided into fifteen birth cohorts from before 1894 up to 1964. In all fifteen cohorts, the childless had higher scores than those with children. The detailed results are shown in Table 5.3.

These results are corroborated by studies of the relation between educational level and childlessness. Educational level is correlated with intelligence at a magnitude of around .6 (Eysenck, 1979) and can be taken as an approximate proxy for intelligence. Studies on the relationship between educational level and childlessness based on large nationally representative samples in Britain, Canada, Norway, Singapore and the United States are shown in Table 5.4. Observe that in all five countries childlessness increases with education level. Observe also that in Britain, for which there are data for both sexes, the educational differentials for childlessness are more pronounced for females than for males, indicating greater dysgenic fertility among women than among men. Given the

Table 5.4
Percentages of Childless in Relation to Educational Level in Various Countries

Country	Date of Birth	Sex	Age	Sample Size	Educational Level				Reference
					Basic	2	3	Tertiary	
Britain	1946	Male	36	1759	5.6	5.4	11.8	10.2	Kiernan, 1989
		Female	36	1831	4.9	5.9	8.9	11.1	Kiernan, 1989
Canada	1935-50	Female	35-49	2082	5.9	7.7	10.3	18.6	Grindstaff, Balakrishan & Dewit, 1991
Norway	1935-54	Female	35+	7789	8.3	9.8	13.0	17.7	Kravdal, 1992
Singapore	1910-40	Female	40-70	Census	4.8	10.3	8.4	13.4	Singapore census, 1980
United States	1946-55	Female White	35-44	16029	8.5	13.1	20.0	27.4	Bachu, 1991
United States	1946-55	Female Black	35-44	2251	15.3	10.5	16.9	24.8	Bachu, 1991
United States	1946-55	Female Hispanic	35-44	1383	7.5	8.4	10.4	18.3	Bachu, 1991

strong association between level of education and intelligence, and also between level of education and childlessness, it is inconceivable that, in the light of the data shown in Table 5.4, the childless can be predominantly those with low intelligence. On the contrary, the childless are predominantly those with high intelligence. This means that the studies showing negative associations between intelligence and number of siblings underestimated the rate of intelligence decline because they omitted the childless.

The explanation for the difference between these results and the conclusions of Higgins, Reed and Reed (1962) and Bajema (1963) is that the latter two studies were concerned with the low fertility of mentally retarded men with IQs below 70. These men probably do have low fertility, but as they only constitute approximately 2.7 percent of the population, they are too few in number to affect the generally negative association between intelligence and fertility. This is evident not only in IQ by sibling-size studies, but also in IQ by fertility studies, as we shall see in Chapters 6 and 7.

5. DOES LARGE FAMILY SIZE ADVERSELY AFFECT IQ?

The second objection that has been raised against the method of calculating the decline in intelligence from the negative association between intelligence and number of siblings is that this association is purely an entirely environmental effect. One of those who has advanced this criticism is Blake (1989). The argument is that in larger families parents have less time to devote to each child and this acts as an environmental depressant on the development of the child's intelligence. Conversely, in small families parents can devote more time to each child, and this enhances the children's intelligence. This position is based on the confluence model of Zajonc and Marcus (1975) which claims that children's intelligence is a function of the amount of attention they receive from adults, and this is typically less in larger families.

There are three arguments against the confluence theory. First, the negative association between number of siblings and intelligence is not present among only children as compared with those in two- and three-child families. According to the theory, only children should receive more parental attention than those in two- and three-child families and should have higher IQs. A number of studies have shown that this is not so and that only children have lower IQs than those from two- and three-child families. This was found in the huge Dutch study of nearly 400,000 military conscripts reported by Belmont and Marolla (1973) where only children had an IQ about five IQ points lower than those in two- and three-child families. This was also found among eleventh grade American children in a study by Breland (1974). In the 1947 Scottish Council (1949) study and in a large French study reported by Vallot (1973) there was no difference between the IQs of only children and those in two-child families. These results are difficult for the confluence theory. Probably the explanation is that most intelligent parents typically plan two- or three-child families. Larger families are frequently unplanned by less intelligent parents. A number of only children are also unplanned by unmarried women who come predominantly from the lower IQ bands, as shown by Herrnstein and Murray (1994). Thus, the failure of the confluence model to hold for family sizes of one to three children suggests that a significant proportion of the negative relationship between family size and intelligence is a genetic effect.

A second argument against the confluence model is that in economically less developed countries there is no relationship between children's IQs and their number of siblings. For instance, it was shown by Ho (1979) that in Hong Kong in the 1970s, the IQs of children in 3–7 child families were higher than in 1–2 child families. This has also been found among Vietnamese refugees in the United States in the late 1980s, where number of siblings was positively associated with educational attainment in mathematics, a good proxy for intelligence—which regrettably was not measured in the study (Caplan, Choy and Whitmore, 1992).

A third problem for the confluence theory is that the negative relationship between intelligence and family size does not hold for adopted children. Scarr and Weinberg (1978) found that among 237 children reared by their natural parents, the usual negative correlation—in this case −.21—was present between family size and IQ, but among 150 adopted children, the correlation was negligible (−.05) and not statistically significant. Clearly, if the negative correlation between intelligence and family size is an environmental effect, it should hold equally for adopted and biological children. But it doesn't. These three failures of the confluence model make it doubtful whether large family size as such is an environmental depressant on children's intelligence. Two critical examinations of the confluence theory of the supposed adverse environmental effects of large family size on children's intelligence by Galbraith (1982) and Retherford and Sewell (1991) both reject the theory as untenable. However, even if the theory contains an element of truth, the supposed adverse environmental effect of large family size would not necessarily be the sole factor responsible for the negative association between number of siblings and IQ. The association could arise partly from the supposed environmental depressant effect and partly from the tendency of more intelligent parents to have fewer children. As long as there is some genetic component responsible for the negative association between number of siblings and intelligence, the association must entail dysgenic fertility.

6. THE DECLINE OF INTELLIGENCE IN BRITAIN, 1955–1980

After 1950 no further attempts were made to estimate the decline of intelligence from the inverse association between intelligence and number of siblings. This was partly because people began to lose interest in this question, and partly because the method was thought to be invalid since childless adults were not taken into account and the inverse association might arise environmentally from the adverse impact of large family size on children's IQs. We have seen that these two criticisms are themselves invalid and that the inverse relationship between IQ and number of siblings can be used for an approximate calculation of the decline of intelligence.

This clears the way for a revival of the method. Accordingly, in 1993 I carried out an investigation of this issue in Britain in collaboration with my colleague Erasmus Harland and our research assistant Lucy Greene. The study involved testing the intelligence of 517 secondary schoolchildren with a mean age of 13.10 years in a city in Northeast England. The test used was the Progressive Matrices. Mean IQs in relation to family size were calculated and are shown in Table 5.5. Notice that the mean IQ is highest in the children with no siblings and declines in more or less linear fashion as the number of siblings increases. The correlation between IQ and the number of siblings is −.18, statistically significant at the 1 percent level but rather smaller than the negative correlations obtained in Britain in the period 1920–1950 which averaged around −.25.

The method used by Burt (1946) and Thomson (1947) is used to estimate the magnitude of the decline in intelligence from these data. The mean IQ of 96.0 of the sample is taken as a measure of the mean IQ of the parents. To calculate the mean IQ of all the children of this sample of parents, the mean IQs of each family size have to be weighted by the number of siblings and the number of families of each family size. This gives a mean IQ of 95.2 for all the children of the parents. Thus, for this generation the mean IQ has declined by 0.8 IQ points. The adolescents in the sample were born in 1980, and it can reasonably be assumed that their parents were born on the average around 1955, assuming a generation length of 25 years. Thus, our result provides an estimate of the decline in intelligence for a British birth cohort born in 1980 as compared with their parents born around 1955.

There are two principal points of interest in this result. First, the magnitude of the generational decline in intelligence in Britain has evidently slackened in the second half of the twentieth century to 0.8 IQ points per generation as compared with the 2.0 IQ points per generation, calculated by Cattell, Burt and Thomson for the 1920s and 1930s. This result is consistent with several other types of evidence which also point to a reduction in dysgenic fertility in the second half of the twentieth century. The most notable of these are the socioeconomic-status fertility differentials, which have also slackened in the second half of the century, as shown in Chapter 10. The evidence hangs together consistently in suggesting strong dysgenic fertility in the first half of the twentieth century followed by weaker dysgenic fertility in the second half.

A second point of interest in our result indicating a generational decline of 0.8 IQ points in Britain for the second half of the twentieth century is that it is closely similar to calculations of the rate of intelligence decline in the United States, calculated by the different method of correlating the intelligence of adults with their number of children. When we examine studies of this kind in Chapter 6, we shall see that Retherford and Sewell (1988) estimate a decline of intelli-

Table 5.5
Mean IQs of British Children in 1993 Analyzed by Their Number of Siblings

	0	1	2	3	4	5+	Total
Number	23	228	171	59	25	11	517
Mean IQ	98.9	97.4	95.5	95.7	87.4	89.6	96.0

gence in the United States at 0.81 IQ points per generation, virtually the same figure as we obtained in Britain.

7. ESTIMATING THE DECLINE OF GENOTYPIC INTELLIGENCE

It is important to note that the rate of decline of intelligence estimated from the inverse association between intelligence and number of siblings is the notional decline of phenotypic (measured) intelligence, that is, the decline in measured intelligence assuming that environmental factors are constant between the two generations. Phenotypic intelligence is a product of genotypic intelligence and environmental influences. What the eugenicists were concerned about was the possible decline in genotypic intelligence, that is, the genetic quality of the population. The problem of measuring the decline of genotypic intelligence is to isolate the genetic component in the decline.

People like Cattell, Burt and Thomson were aware of this. Their position was that genetic factors are the more important determinants of intelligence and, therefore, that the decline was largely genetic, but they were not able to put a precise figure on the magnitude of the genetic decline because the concept of heritability had not been satisfactorily quantified at this time. They therefore confined themselves to the conclusion that the decline was largely genetic.

Later in the century techniques were developed for quantifying the genetic and environmental contributions to intelligence and it became possible to estimate the magnitude of the genetic component in the decline. The problem was to measure heritability, the contribution of genetic factors to the variance in intelligence. Once this was done, the decline in genotypic intelligence is obtained by multiplying the decline by the heritability.

For this purpose we need an estimate of the heritability of intelligence. The two most commonly used methods are correlations between pairs of identical twins reared in different families, and the difference between the correlations of pairs of identical and nonidentical twins reared in the same family. The correlation between the IQs of pairs of identical twins brought up in different families provides a direct measure of heritability. The reason for this is that the twins are genetically identical, so any differences between them must be caused by the different environments. If there is no difference between them, the environment must be having no effect, and the correlation will be 1.0, indicating complete heritability. If the correlation is a little less than 1.0, heritability must be high. In fact, the correlation from all existing studies of adults is .72 (Bouchard, 1993, p. 58). This correlation needs correction for the reliability of the tests, which Bouchard estimates at .9. Making this correction raises the heritability to .80. This is why most leading authorities, such as Jensen (1972, p. 294) and Eysenck (1979, p. 102), have concluded that the heritability of intelligence is about .80 or 80 percent.

The alternative method is to use the difference between identical and nonidentical twins reared in the same family. Identical twins are more similar than nonidenticals because identicals have both their genes and their environment in common, whereas nonidenticals have only half their genes and their environment in common. A simple method for estimating heritability was proposed by Falconer (1960) and consists of doubling the difference between the correlations of identical and nonidentical twins. All the data on these correlations has been summarized for adult pairs by Bouchard (1993, p. 58), who calculates that the correlations are .88 for identicals and .51 for nonidenticals. These correlations also need correction for test unreliability; using the reliability figure of .9 raises them to .98 and .56, respectively. The difference between the two correlations is .42 and double the difference is .84. This is the measure of heritability obtained by this method. For our own purposes, I shall average the two heritability estimates of .80 and .84 to give a single best estimate of .82.

This figure is somewhat higher than some authorities have used. For instance, Mackintosh and Mascie-Taylor (1984) propose that the heritability of intelligence lies somewhere between .25 and .75, and Herrnstein and Murray (1994) that it lies between .40 and .80. The reason for these lower figures is that they are derived largely from children. Heritabilities are lower among children because the environment has temporary effects which wash out by adulthood. Thus, heritabilities derived from the identical-nonidentical differences increase from .40 among 4 to 6-year-olds, to .54 among 6 to 16-year-olds to .84 among adults (Bouchard, 1993, p. 58). For this reason it is best to calculate heritability from data on adults.

The application of a heritability of .82 to the decline in phenotypic intelligence per generation calculated for Britain in the 1920s–1940s is that the genotypic decline is obtained by multiplying the phenotypic decline of two IQ points by the heritability. This gives a figure of 1.64 IQ points per generation for the decline in genotypic intelligence. In the second half of the twentieth century, our estimate of a phenotypic decline of 0.8 IQ points per generation indicates a genotypic decline of .66 IQ points.

Perhaps we should conclude by noting that although some readers may consider a heritability of .82 for intelligence as too high, the precise figure adopted does not make a great deal of difference to the magnitude of the intelligence decline. If heritability is set at .6 as preferred by Herrnstein and Murray, the genotypic decline becomes 1.20 IQ points per generation for the first half of the century, instead of 1.64. The main point is that intelligence in Western nations is in decline, whatever the precise rate at which it is taking place.

8. CONCLUSIONS

There is little doubt that those who worked on the problem of dysgenic fertility in the first half of the century were right in their belief that the inverse association between intelligence of and number of siblings indicated that the more intelligent parents of their samples were having fewer children than the less intelligent, and consequently that the intelligence of the population was in decline. The objection that the method omitted the childless was certainly invalid. The objection that family size acted as a purely environmental depressant on intelligence was also incorrect, although this may have introduced some element of error into the estimates. Nevertheless, any overestimate of the environmental depressant effect on the intelligence of members from large families is probably approximately counterbalanced by the underestimate caused by the omission of childless couples. For these reasons, the estimate of a decline of genotypic intelligence of 1.64 IQ points derived from the British sibling-size studies of the 1920s–1940s can be regarded as approximately accurate. The magnitude of the decline in the United States for the 1920s, calculated by Lentz at 4.0 IQ points per generation and therefore twice as great as the decline obtained in Britain at this period, looks a bit high—possibly because of a sampling error in his particular sample.

We can use the British studies on the inverse relation between children's IQs and their number of siblings to estimate the impact of dysgenic fertility on the genotypic intelligence of the population over a period of approximately 90 years, from 1890 to 1980. For the birth cohort of 1890, we can use the study carried out by Sutherland and Thomson (1926) in 1925 on 10 and 11-year-olds, from which Thomson (1947) calculated a decline from parents to children of 2.1 IQ points. These children would have been born about 1915, and their parents born, on average, about 1890, on the assumption of an average generational length of 25 years. This gives an IQ decline of 2.1 IQ points for the generation born around 1890 to that born in 1915.

The next two studies by Cattell and by Roberts, Norman and Griffiths were for children born about 1925 and whose parents were born about 1900. The generational declines in the two studies were 2.2 and 2.04 IQ points, respectively—virtually the same as that for the 1890 parental cohort. The next major study was that of approximately 70,000 Scottish 11-year-olds tested in 1947. The correlation between their IQs and number of siblings was −.28, somewhat higher than the correlations of the 1920s–1930s of −.22 to −.23. We can therefore assume that the rate of decline of approximately two IQ points per generation persisted for this cohort whose parents would have been born around 1915.

Our study of the parental birth cohort of approximately 1955 showed a slackening of the rate of decline to .8 IQ points per generation. We are missing data for the parental birth cohort born around 1940, but it seems reasonable to assume that this would have been about midway between the parental birth cohorts of 1915 and 1955, i.e., 1.4 IQ points. Working on this assumption, we have estimates of the decline of intelligence for four parental generations born around 1890, 1915, 1940 and 1955 of 2.0, 2.0, 1.4 and .8 IQ points. This gives a decline of intelligence of 6.2 IQ points over the 90-year period and, assuming a heritability of .82, gives a genotypic decline of 5.0 IQ points.

Chapter 6

Intelligence and Fertility in the United States

1. Higgins, Reed and Reed, 1962. 2. Bajema, 1963 and 1968. 3. Waller, 1971. 4. Osborne, 1975. 5. Vining, 1982 and 1995. 6. Van Court and Bean, 1985. 7. Retherford and Sewell, 1988. 8. Herrnstein and Murray, 1994. 9. Dysgenic Fertility among Whites, Blacks and Hispanics. 10. Dysgenic Fertility in Males and Females. 11. Conclusions.

Even if it were basically sound, the method of using the negative correlation between intelligence and the number of siblings to calculate the extent of dysgenic fertility entailed assumptions that can be questioned. A more straightforward method for determining whether genotypic intelligence is declining is to examine the relationship between the intelligence of adults and their number of children. If the relationship is negative, genotypic intelligence must be deteriorating. It was not until the second half of the twentieth century that studies based on adequate samples began to appear. We look first at the American studies and turn in the next chapter to studies in Europe.

1. HIGGINS, REED AND REED, 1962

The decade from 1962 to 1971 saw the appearance of the first adequately sampled American studies of the relationship between intelligence and number of children. The first of these was carried out by Higgins, Reed and Reed (1962). It was based on a sample in Minnesota which appears to have been born between 1910 and 1928. The initial sample consisted of approximately 2,032 mothers and fathers, and 2,039 of their children, for all of whom there were intelligence test results. The correlations between intelligence and number of children were negative for both fathers (−0.08) and mothers (−0.11), indicating that more intelligent adults had fewer children. Both cor-

Table 6.1
Fertility for Six IQ Bands among Married People with Children and in the Total Sample

	Married with children		Total Sample	
IQ Band	N	N. Children	N.	N. Children
0-70	73	3.81	103	2.16
71-85	180	2.98	208	2.39
86-100	597	2.65	583	2.16
101-115	860	2.68	778	2.26
116-130	287	2.70	269	2.45
131+	35	2.94	25	2.96
Total	2,032	2.75	1,966	2.27

Source: Higgins, Reed and Reed (1962).

relations are statistically significant, indicating significant dysgenic fertility for both males and females.

The data in this study are shown in Table 6.1, which gives a breakdown of intelligence into six IQ bands running from below 70 to 130 +, and their average number of children. Notice that the greatest number of children were produced by those in the lowest IQ band of below 70 (3.81 children) and the next greatest by those in the 71–85 IQ range (2.98 children). The number of children falls to 2.65–2.70 in the middle intelligence range and then increases in the highest IQ group of 130+ (2.94 children). However, there were very few of these, and it is obvious from inspection that the overall relationship between intelligence and fertility is negative. There is, in addition, some tendency towards a U-shaped relationship between intelligence and fertility with lowest fertility in the middle of the range and highest fertility at the extremes.

The authors were concerned that their sample was confined to people with children. It was not, therefore, a representative sample of the population, which includes people who do not have children. There are two groups missing from the sample, namely, those who were married but did not have children, and those who were unmarried and did not have children. To overcome this problem, the authors first added the married but childless siblings of the parental sample to the group. When they did this, the association between intelligence and fertility remained negative. They then added the unmarried siblings of the parental sample. This time the relationship between intelligence and fertility became slightly positive. The results are shown in the fourth and fifth columns of Table 6.1. Note that those in the lowest IQ band of 0–70 have a lower than average number of children (2.16) and that the greatest number of children occur in the two highest IQ bands of 116–130 and 130+. The authors did not give the correlation between intelligence and number of children, but they asserted that it was positive.

There is, however, an odd feature of the data: when married but childless siblings and unmarried siblings were added to the initial sample, the total number became fewer and not, as would be expected, greater. The authors do not explain this curiosity. Nevertheless, they claimed that their study showed that the overall relationship between intelligence and fertility is slightly positive.

The study is open to criticism on three grounds. First, there is the numerical anomaly that adding the childless to the original sample had reduced the sample size, whereas it should have increased it. There are obviously errors in the data here. Second, the essential point of the authors' argument is that the childless have below-average intelligence. This again raises the question, Who are the childless? In Chapter 5, we saw that there is a mass of evidence that the childless have above-average intelligence. The authors must have found a seriously atypical group of childless adults with below-average intelligence. Third, there is a problem about the representativeness of the sample. It was drawn from a collection of pedigrees of mental retardates built up by the Institute for Human Genetics at the University of Minnesota. These retardates had a number of relatives of normal intelligence, and they formed the initial sample of the study. A sample obtained in this way is unlikely to be representative of the population. This suspicion is strengthened by the distribution of intelligence shown in the fourth column of the table. A representative sample of 1,966 individuals would have 45 with IQs of 70 and below (2.3 percent) and the same number with IQs of 130 and above. This sample has more than twice as many with IQs below 70 as it should have, and only about half the number of those with IQs above 130. The overall verdict on this study has to be that it contains too many anomalies to provide any clear indication of the direction of the relationship between intelligence and fertility.

2. BAJEMA, 1963 AND 1968

The next studies to tackle this question were carried out by Bajema (1963, 1968). The first of these used a sample of 979 native-born white individuals born in 1916 and 1917 and brought up in Kalamazoo, Michigan. Their intelligence was tested in 1927 and 1928 when they were 11 years old. They were followed up in the year 1951 when they were aged 45 and 46. Their number of children, marital status, place of residence and date of death, if deceased, were recorded. The sample was divided into five IQ bands. Table 6.2 gives the mean

Table 6.2
Intelligence and Fertility in the Kalamazoo Sample

IQ Band	N	Mean N. Children	Percent Childless	Percent Unmarried
120+	82	2.60	13.41	6.10
105-119	282	2.24	17.02	4.96
95-104	318	2.02	22.01	6.60
80-94	267	2.46	22.47	5.62
69-79	30	1.50	30.00	10.00

Source: Bajema (1963).

number of children for each IQ band, the percentage childless and the percentage unmarried.

It will be noted that the group with the highest intelligence of 120+ had the greatest number of children (2.60), while those with the lowest intelligence of 69 to 79 had the lowest number of children (1.50). Note also that the relation between intelligence and fertility is not linear. Fertility declines with intelligence in the three highest IQ bands, but it increases again among the 80–94 IQ group. If the group with IQs between 69 and 79 is discounted because of the small number, the overall relationship between IQ and fertility becomes U-shaped; those with the highest and lowest IQs having greater fertility than in the middle range, as in the Higgins, Reed and Reed study. The overall correlation between intelligence and fertility is 0.05, indicating slight eugenic fertility.

The sample was analyzed further to ascertain the relationship between sibling size and intelligence. This was found to be −0.26 and is therefore consistent with a large number of studies summarized in section 1 of Chapter 5. Bajema noted that it is possible to have a sample where the correlation between intelligence and sibling size is negative and the correlation between intelligence and fertility is positive. He argued that the explanation for this difference is that those who are childless do not appear in data on the relation of intelligence to number of siblings, and that childlessness is more prevalent in the lower IQ bands. This is shown in column four of the table, in which the percentage childless increases steadily from 13.41 in the IQ band 120+ to 30.00 in the IQ band 69–79. To some degree this represents intelligence differences in marriage rates, as shown in the last column of the table, but the major factor appears to be infertility among lower IQ bands rather than the lower incidence of marriage.

The principal problem about the sample is it representativeness. The sample was urban, white and Protestant. The omission of a rural population will have had some distorting effect because rural populations invariably have higher fertility and generally have lower intelligence than urban populations. For instance, in a sample born about the same time as the Kalamazoo sample, the fertility of American farm wives was 2.7 as compared with 2.1 for all American wives (Freedman, Whelpton and Campbell, 1959). At the same time, rural populations had lower average intelligence levels in the United States at about the relevant time. McNemar (1942) reported a 12.2 IQ point difference between the urban and rural populations in the standardization sample of the Terman-Merrill test. Hence, if the Kalamazoo sample had been representative of the population of the United States by including a rural group, the positive association between intelligence and fertility would have turned neutral or negative. A further problem with the sample is the inverse association between intelligence and childlessness. As we have seen, the weight of the evidence indicates that this relationship is positive. Finally, the fact that there was a negative correlation between intelligence and number of siblings and a positive correlation between intelligence and fertility does not really represent an anomaly. The negative correlation between intelligence and number of siblings provides a measure of the decline of intelligence from the sample's parents to the sample. The small positive correlation between intelligence and fertility indicates eugenic fertility in the next generation, from the sample to the sample's children. There almost certainly was a reduction in dysgenic fertility over the course of these two generations, as shown in more detail in Chapter 10, so the two correlations are not necessarily inconsistent.

Bajema's (1968) second investigation was based on a sample drawn from the Third Harvard Growth Study and consisted of white native-born Americans. The results are shown in Table 6.3. The correlation between intelligence and fertility is .04. However, as with his first study, there are problems about the representativeness of the sample. The average number of children for the sample was only 2.14, whereas the mean fertility for representative samples of American whites born between 1912 and 1930 was approximately 2.7 (Van Court and Bean, 1985), suggesting that the sample was unrepresentative. One reason for this was that the sample was entirely urban and therefore omitted the lower IQ/ higher fertility rural population. Because of these sampling inadequacies, the two Bajema studies do not provide satisfactory evidence on the issue of whether fertility in the United States at that time was dysgenic.

3. WALLER, 1971

The remaining study of the 1962–1971 period was carried out by Waller (1971). His sample was an updated group obtained from the Minnesota Mental

Table 6.3
Number of Individuals Broken Down into IQ Bands and Their Average Number of Children

	IQ Band						
	0-55	56-70	71-85	86-100	101-115	116-130	130+
Number	3	19	216	506	510	201	78
Children	2.00	1.21	2.00	2.21	2.18	1.99	2.40

Source: Bajema (1968).

Retardation study analyzed some ten years previously by Higgins, Reed and Reed (1962). The results are shown in Table 6.4. This gives a breakdown of the samples into seven IQ bands and the mean number of children per individual in each band. It can be seen that there is a positive association between intelligence and fertility, the value of the correlation being .11.

The sample in this study has the same problems regarding its representativeness as the original Higgins, Reed and Reed sample, on which it was based, and these raise doubts about the reliability of the results.

Nevertheless, by 1971, four studies had appeared on the relationship between intelligence and fertility in the United States; in three the relationship was positive, while in the fourth, Bajema's second study, the relationship was zero. Although the samples were not fully representative of the population, the consistency of the results was to some degree impressive and led most of those concerned with this issue in the 1970s to conclude that American birthrates were slightly eugenic for cohorts born from 1912–1928 onward. It seemed that the pessimism of the eugenicists of the first half of the century was unwarranted and that a mild degree of optimism was justified. This conclusion was to receive a setback when further studies of the relation between intelligence and fertility were to appear over the years 1975–1994.

4. OSBORNE, 1975

The conclusion that the relationship between intelligence and fertility is positive or neutral was first challenged by a study carried out by Osborne (1975) in the state of Georgia. In 1971 Osborne obtained data on the IQs of all children aged 10 to 14 in Georgia, numbering approximately a quarter of a million. He then calculated the average IQs of children in each of 159 counties. He also

Table 6.4
Number of Individuals Broken Down into IQ Bands and Their Average Number of Children

	IQ Band						
	0-55	56-70	71-85	86-100	101-115	116-130	130+
Number	3	13	41	209	296	133	18
Children	0	2.92	3.12	3.42	3.77	3.33	4.06

Source: Waller (1971).

obtained the fertility ratio of these counties, comprising the number of children per 1,000 women aged 15 to 49. The fertility ratio included the number of children of unmarried women and of married women with no children. Six correlations were computed between average fertility ratios and IQs, for verbal and non-verbal tests for three age groups. The correlations ranged from −0.43 to −0.54 and averaged −0.49, which is highly statistically significant on 159 population units. The population in this study included both blacks and whites, who comprised 26 percent and 74 percent of the population, respectively.

This study evidently shows a strong dysgenic trend for intelligence and fertility, in contrast to the four initial studies. Osborne's study has merits as compared with the initial studies because the sample was the entire population of a state, and the fertility ratio was based on all women, including the single and the childless. Furthermore, the sample includes blacks as well as whites. In all these respects the sample is superior to those of the four initial studies, which were less representative and did not include blacks. The presence of both blacks and whites in a sample examining the relation between intelligence and fertility is likely to turn the association negative, because blacks have lower IQs and greater fertility than whites.

There are some limitations to this study: First, it was based only on Georgia; second, the sample combines blacks and whites, and it would have been better to have the correlations between intelligence and fertility for blacks and whites separately; and third, the correlations were calculated from population units rather than from individuals, which generally inflates the size of the correlations. Nevertheless, the negative correlations between intelligence and fertility were high and the results place a question mark over the positive correlations obtained in the earlier studies.

5. VINING, 1982 AND 1995

The negative correlation between fertility and intelligence found by Osborne on group data was confirmed a few years later on individuals by Vining (1982, 1995). The initial study was based on data for 5,172 males and 5,097 females born between 1942 and 1954, for whom intelligence test results and other information were obtained between 1966 and 1968. Fertility data were obtained between 1976 and 1978, when the sample was aged 25 to 34. The data came from the National Longitudinal Study of Labor Market Experience. This was a nationally representative sample including both blacks and whites, except for the exclusion of high-school dropouts.

For men, fertility was measured by taking the number of children living with their fathers at the time of the survey. Vining regarded this as unsatisfactory for blacks because of the large number of unstable unions but as an adequate, although imperfect, measure for whites. For women, information was available for the number of children they had, the total number they expected to have and the number considered ideal for a family.

Intelligence was assessed from scores on various tests obtained for those who had completed ninth grade. The scores came from about 30 different tests which were transformed to a common scale with a mean of 100 and standard deviation of 15. Unfortunately, information on both intelligence and fertility was available for only 61 percent of white males and females, 27 percent of black males and 35 percent of black females, which reduces the sample size quite considerably and almost certainly distorts its representativeness.

The three groups were divided into six IQ bands running from below 70 to 130+ and number of children calculated for each IQ band. These results are shown for white women, black women and white men in the first three rows of Table 6.5. Notice that for three groups, the high IQ bands tend to have the fewest children. This is confirmed by the negative correlation coefficients for IQ and number of children shown in the right-hand column. The negative correlations are all statistically significant and are greatest for black females and smallest for white males.

Since the individuals in this sample were aged 25 to 34 at the time of the study, many of them would not have completed their fertility. In an attempt to overcome this problem and get an estimate of what their completed fertility was likely to be, the women were asked about their expected number of children. These expectations are given for white and black women in the fourth and fifth rows of the table. The negative correlations between IQ and expected number of children are smaller than the correlations for number of children already born. This reflects the fact that more intelligent women tend to delay their childbearing. Nevertheless the correlations are still significantly negative, although the expectations will not necessarily be precisely fulfilled. Vining also gives results for the ideal number of children for the different IQ groups, and these are shown for white and black women in the sixth and seventh rows of the table. The

Table 6.5

Fertility among Different IQ Bands Born 1942–1954 Showing Negative Correlations between IQ and Fertility

				IQ Band							
Group	No.	Age	Measure	0 - 70	71 - 85	86 - 100	101 - 115	116 - 130	130+	All	Corr.
White women	2066	25-34	N. children	1.59	1.68	1.76	1.44	1.15	.92	1.46	- .177*
Black women	473	25-34	N. children	2.60	2.12	1.79	1.63	1.20	.00	1.94	- .202*
White men	1993	25-34	N. children	1.17	1.30	1.29	1.19	.84	.45	1.14	- .140*
White women	2039	25-34	Expected N. children	2.31	2.16	2.30	2.14	2.03	1.93	2.15	- .083*
Black women	465	25-34	Expected N. children	3.20	2.75	2.36	2.25	2.30	2.00	2.56	- .171*
White women	2049	25-34	Ideal N. children	2.41	2.39	2.34	2.33	2.31	2.49	2.34	- .031
Black women	462	25-34	Ideal N. children	2.94	2.75	2.74	2.43	2.40	2.00	2.70	- .086
White women	1839	35-44	N. children	1.80	2.00	2.14	1.92	1.78	1.77	1.94	- .062*
Black women	378	35-44	N. children	3.04	2.74	2.06	1.93	2.00	- - - -	2.31	- .226*

Note: Asterisks denote statistically significant correlations at the 5 percent level or less.
Source: Vining (1982, 1995).

correlations for these are not statistically significant. Hence, Vining's results showed a negative relationship between intelligence and fertility, while the earlier studies reviewed in the preceding sections found positive relationships. He suggested that during periods of declining fertility, the intelligence-fertility relationship is negative, while in periods of rising fertility it becomes positive. Thus, the relationship was dysgenic among birth cohorts born from around 1850 to 1910—a period of falling fertility—turned eugenic among those born around 1910 to 1940 when fertility began to increase (shown in the first four studies), and then turned dysgenic again in his own cohort born between 1942 and 1954.

Vining's results contain a number of points of interest. First, they show the presence of dysgenic fertility for both whites and blacks. Second, dysgenic fertility is greater for blacks than for whites. Third, Vining suggests that the results underestimate the true extent of dysgenic fertility because of the exclusion of high-school dropouts who probably had low IQs and high average fertility, and this is almost certainly correct. Fourth, most women expect to have fewer children than they consider ideal. Fifth, this is particularly true for high-IQ white women with an expectation of 1.93 children and an ideal of 2.49. However, it is not true for high-IQ black women, whose expected and ideal number of children are the same at 2.00. Sixth, black women with IQs below 70 expect to have more children (3.20) than they consider ideal (2.94), apparently reflecting resignation to the probability of having more children than they would wish. Finally, the correlations between intelligence and ideal number of children are effectively zero, and this is an important finding because it indicates that in an ideal world there would be no dysgenic fertility. However, there is dysgenic fertility in the real world, largely because the more intelligent white women expect to have significantly fewer children than they consider ideal, while the least intelligent black women expect to have significantly more children than they consider ideal.

The most satisfactory data in the relationship between intelligence and fertility comes from individuals who have completed their childbearing. Ideally, this means those aged 45 and over. Vining's initial sample of 25 to 34-year-olds falls short of this ideal, but he subsequently published data on the fertility of his female sample at the ages between 35 and 44 (Vining, 1995), and for practical purposes this is close to completed fertility. The results are shown for white and black women in the last two rows of Table 6.5. The correlations between intelligence and fertility are still significantly negative. However, the change in the size of the negative correlations differs for whites and blacks. For white women, the negative correlation falls from 25 to 34-year-olds to 35 to 44-year-olds, from −0.177 to −0.062, reflecting the greater tendency of more intelligent women to have their children in their thirties. Among black women, the negative correlation increases from −0.202 to −0.226.

The principal problem with this study concerns the sampling. First, it excluded high-school dropouts, who were about 14 percent of whites and 26 percent of blacks at this time; second, only 61 percent of whites and 35 percent of blacks in the sample provided data for the 25 to 34-year-olds; and third, the sample was reduced further for the 35 to 44-year-olds. It is difficult to estimate by how much this would have distorted the results.

6. VAN COURT AND BEAN, 1985

The next study to appear on fertility and intelligence was to confirm Vining's results. Van Court and Bean (1985) obtained a sample of 12,120 American adults for whom data on intelligence, number of siblings and number of children were available. Intelligence was assessed by a ten-item vocabulary test, a short but adequate measure. The data were analyzed for fifteen birth cohorts spanning the years of birth from 1890–1894 to 1960–1964. In all the cohorts, the correlations between intelligence and number of children were negative. The results are shown in Table 6.6 which gives the birth cohorts; the correlations between intelligence and fertility for the sample; the correlations for whites only; and correlations between intelligence and number of siblings. To provide an overall picture, the correlations have been averaged giving each cohort equal weight; these averages are shown in the bottom row of the table.

There are five interesting points in these results. First, the correlations between intelligence and fertility in this sample are consistently negative. The overall correlations are −0.17 for the total sample and −0.16 for whites. Clearly these correlations confirm the results of Osborne and Vining and not the earlier studies showing positive correlations between intelligence and fertility. Second, this discrepancy cannot be resolved by Vining's hypothesis that the negative association between intelligence and fertility which prevailed up to the 1910 birth cohort turned positive in the birth cohorts 1910–1940, and then turned negative again. There are no cohorts where the correlation is positive, although among the whites the correlation does fall to zero for the 1925–1929 birth cohort and to −0.04 for the 1930–1934 cohort, suggesting a slackening of the dysgenic trend among these cohorts in line with Vining's theory.

Third, the correlations between intelligence and fertility fall from the cohorts born before 1909 (mean correlation = −0.21) to the cohorts born 1910–1944 (mean correlation = −0.13). This indicates a weakening of the dysgenic trend as the twentieth century progressed, probably because knowledge and practice of birth control spread among the poorer, less educated and less intelligent classes from around 1910 onward. The increase in the size of the negative correlations from the birth cohort 1935–1939 onward probably reflects the fact that these cohorts had not completed their fertility at the time of the surveys. The more intelligent tend to delay their childbearing; hence, correlations between intelligence and fertility in samples who have not completed their fertility exaggerate the dysgenic trend.

Fourth, the correlations between intelligence and number of siblings are all negative and average −0.29, in line with the numerous other studies reviewed in section 1 of Chapter 5. This result is useful because it shows that on the same

Table 6.6
Negative Correlations between IQ and Fertility in Fifteen Cohorts in the United States

		Correlations		
		Total Sample	Whites	N. Siblings
Cohort	Date of Birth	r	r	r
1	1890-1894	-.06	-.04	-.12
2	1895-1899	-.24	-.20	-.37
3	1900-1904	-.26	-.26	-.34
4	1905-1909	-.19	-.19	-.28
5	1910-1914	-.07	-.09	-.22
6	1915-1919	-.14	-.14	-.35
7	1920-1924	-.13	-.14	-.23
8	1925-1929	-.11	.00	-.27
9	1930-1934	-.04	-.04	-.32
10	1935-1939	-.18	-.18	-.34
11	1940-1944	-.16	-.19	-.35
12	1945-1949	-.27	-.27	-.37
13	1950-1954	-.24	-.24	-.30
14	1955-1959	-.25	-.23	-.31
15	1960-1964	-.23	-.25	-.09
Total	1890-1964	-.17	-.16	-.29

Source: Van Court and Bean (1985).

sample, the two methods for assessing dysgenic trends—the measurement of IQ by fertility and IQ by sibling size—produce negative results and are therefore consistent. This lends support to the validity of the method adopted by eugenicists in the first half of the century for measuring dysgenic fertility from IQ by sibling-size data. It is true that in the Van Court and Bean data, the IQ by sibling-size correlation (−0.29) is somewhat higher than the IQ by fertility data (−0.17). However, there is no reason why these two correlations should be precisely the same. The IQ by sibling-size correlation is a measure of the strength of dysgenic fertility from the parents of the sample to the sample itself.

Table 6.7
Correlations between Intelligence with Actual Number and Planned Number of Children and t Values for Statistical Significance of Differences between the Two Correlations

Date of Birth	Number of Children		Difference
	Actual	Planned	t
1945-49	-.27	-.23	9.40***
1950-54	-.24	-.18	22.00***
1955-59	-.25	-.14	25.59***
1960-64	-.23	-.21	3.07**

Note: ** and *** denote statistical significance at the 1 and .1 percent levels, respectively.
Source: Van Court (1985).

The IQ by fertility correlation is a measure of the strength of dysgenic fertility from the sample generation to the sample's children. The fact that the first correlation is higher than the second means that the dysgenic fertility of the early generations has slackened somewhat among the later generations, and this is confirmed by the reduction in the size of the correlations between intelligence and fertility from the cohorts born before 1909 to those born between 1910 and 1944.

Fifth, the correlation between intelligence and fertility is fractionally larger for the total sample (−0.17) than for whites only (−0.16). Because there are relatively few blacks in the sample, the negative correlation must be substantially greater among blacks and can be calculated at −0.35.

Three further results were calculated from this data set which have never been published but are available in Van Court (1985). First, the mean IQs of the childless and of those with children were calculated. In all cohorts the childless had higher average IQs, contrary to the findings of Higgins, Reed and Reed (1962) and Bajema (1963). These results have already been presented in Chapter 5 in Table 5.3. Based on this large nationally representative sample, they cast serious doubt on the claims that the childless have below-average intelligence.

Second, a comparison was made for cohorts 12–15 between the correlation of intelligence and number of children, and the correlation between intelligence and number of planned children. In each case the correlations between intelligence and fertility were significantly lower for planned children than for actual children. The results are shown in Table 6.7. The inference to be drawn from the results is that some of the negative association between intelligence and fertility arises because a greater proportion of births among the less intelligent are unplanned. Hence, the negative correlation between intelligence and fertility

Table 6.8
Number of Children for Men and Women in Five IQ Groups in the United States

	Vocabulary Scores				
	0-1	2-3	4-5	6-7	8-10
Men	2.24	2.05	1.97	1.86	1.79
Women	2.53	2.72	2.45	2.21	1.87

Source: Van Court (1985).

would fall if all births were planned. The size of the reduction is about 25 percent. However, this is probably underestimated in the data because some of the subjects will have rationalized unplanned births and come to think of them as planned.

The third useful analysis consists of a breakdown of fertility in relation to intelligence for men and women separately. The results are shown in Table 6.8. The sample was broken down into five intelligence bands on the basis of their vocabulary scores 0–1, 2–3, and so forth, and the average number of children of men and women are shown for each band. Notice that although the correlations are not given, fertility is evidently more dysgenic for women than for men, in line with several other studies where the sexes are considered separately. Furthermore, those in the lowest IQ bands have the largest number of children, contrary to the earlier results summarized in sections 1–3 of this chapter, claiming that those with low IQs had fewer children than the average.

This study has been criticized by Retherford and Sewell (1988) on three counts. First, they complain that there is no report on the response rates to the surveys. But as this was a quota and probability sample, those who refused to be interviewed would have been replaced by others of the same age group, socioeconomic status, geographical location, race, and so forth, to give a sample representative of the United States so that refusals to respond would not have seriously distorted the representativeness of the sample. Second, they say that there is no information on the validity or reliability of the vocabulary test as a measure of intelligence. This is unfair, because the authors provide adequate justification for both of these. They show that the vocabulary test correlates well with longer intelligence tests and with both educational and occupational achievement. In fact, size of vocabulary is one of the best measures of intelligence available (Jensen and Reynolds, 1982). This covers the validity point, and as for reliability, the authors do in fact provide the figure of 0.79. Third, they say that the survey does not take into account the mortality of siblings and children, but these are so low that they could not have had any significant effect on the results. For these reasons the criticisms of the study are unfounded. This is the most impressive study on dysgenic fertility we have yet seen, being based on a very large nationally representative sample, and the results are unambiguous. Fertility in the United States has been dysgenic over a period of approximately 75 years (1890–1964) spanning three generations.

7. RETHERFORD AND SEWELL, 1988

This study by Retherford and Sewell (1988) was based on a random sample of 10,317 high-school seniors in Wisconsin for whom intelligence test data were collected in 1957. The sample was almost entirely white (98.2 percent, with 0.9 percent black and 0.9 percent other races). The sample was followed up in 1975, when they were approximately 35; a response rate of about 90 percent was secured; and the number of children of the sample were recorded. The sample evidently has some weaknesses. First, it does not include school dropouts who have lower than average IQs, generally below 85, and possibly higher fertility. Second, fertility is not invariably complete by the age of 35. However, the authors make reasonable estimates to correct these omissions.

The relation between the IQs of the sample and fertility is shown for females and males separately in Table 6.9. The sample was divided into ten IQ groups of approximately equal size, and the mean number of children born are shown for the ten groups. It will be seen that for females the trend is clearly dysgenic. Females in the two lowest IQ bands have 2.76 and 2.92 children; among higher IQ bands, the number of children gradually falls, until in the highest band it is 2.29. Among males, the dysgenic trend is not so pronounced but is still present: the five low IQ bands have generally higher fertility than the five high IQ bands, and the lowest fertility occurs in the highest IQ band. For both males and females the negative association between IQ and fertility is statistically significant.

The figures shown in the table were adjusted to take into account the omission of dropouts and of fertility over the age of 35. For dropouts, the authors conclude that fertility is considerably higher among females than among males. When these are added to the data, the authors calculate the selection differential (i.e., the differences between the mean IQ of the sample and that of their children); this is calculated by assuming that on average the mean IQ of the children is the same as that of their parents and weighting the IQ of the parents by their number of children. This gives a selection differential of −1.33 for females and −0.28 for males, and an average of −0.81 for both sexes combined. This means that there has been a theoretical decline in the average IQ of the children of 0.81 IQ points as compared with that of the parents.

As in the studies of intelligence and number of siblings reviewed in Chapter 5, this calculation of the intelligence decline represents the notional decline of phenotypic intelligence, assuming no relevant environmental effects. To calculate the decline in genotypic intelligence, the figure has to be multiplied by the heritability. To do this, Retherford and Sewell adopt a heritability of intelligence

Table 6.9
Mean Number of Children Ever Born by Thirty-Fifth Birthday by IQ Decile and Sex

		Females	Males
IQ Decile	IQ Range	Children	Children
1	67 - 81	2.76	2.36
2	82 - 87	2.92	2.45
3	88 - 92	2.83	2.46
4	93 - 96	2.81	2.50
5	97 - 100	2.73	2.29
6	101 - 103	2.70	2.27
7	104 - 108	2.70	2.26
8	109 - 112	2.55	2.19
9	113 - 120	2.50	2.37
10	121 - 145	2.29	2.07

Source: Retherford and Sewell (1988).

of 0.4, and this gives a genotypic decline of −0.32 IQ points per generation (−0.81 × .4 = −0.32). This figure for the heritability of intelligence is much too low for adults, for whom we estimated in Chapter 5 a heritability of 0.82. Adopting this figure, the decline in genotypic intelligence found in this study is more than doubled to 0.66 IQ points per generation. This is the best estimate available of the decline in genotypic intelligence in the United States for whites born about 1940. Notice that the figure is for whites only because there were virtually no blacks (0.9 percent) in the sample.

8. HERRNSTEIN AND MURRAY, 1994

The latest American evidence to date for intelligence and number of children has been provided by Herrnstein and Murray (1994). They examined the 1990 *Current Population Survey*, a survey of a representative sample of American women, and extracted information on years of education and number of children for the 35–44 age group. This sample was born between the years 1946–1955. Herrnstein and Murray had no direct measure of the sample's intelligence, but

Table 6.10
Number of Children of Women Aged 35–44 Shown in Relation to Their Years of Education and IQs

Years of Education	IQ	Number of Children
16+	111	1.6
13-15	103	1.9
12	95	2.0
0-11	81	2.6
Average	98	2.0

Source: Herrnstein and Murray (1994).

they had data on education levels from which they estimated IQs. Their method was to take the average IQs of those with different levels of education from the National Longitudinal Study of Youth data, and assign these IQs to the women of different educational levels in the Current Population Survey. The results are given in Table 6.10 and show that the best educated and most intelligent women have the fewest children, an average of 1.6. The number of children increase steadily as education and intelligence decline, down to the poorest educated and the least intelligent, who have 2.6 children.

Herrnstein and Murray make a second estimate of the relationship between intelligence and fertility. They consider the educational attainment of all American women who had a baby in the year 1990, taking their data from the National Center for Health Statistics (NCHS). Transforming years of education to IQs, using the NLSY results, they calculate that the average IQ of all women giving birth was 97. The conclusion is that the average baby born in America in 1990 came from a mother with slightly below average intelligence. The result confirms the first estimate that fertility in the United States around the year 1990 was dysgenic.

9. DYSGENIC FERTILITY AMONG WHITES, BLACKS AND HISPANICS

Several of the studies we have reviewed have found greater dysgenic fertility among American blacks as compared with whites, and there is also some evidence for greater dysgenic fertility among American Hispanics. It is now time to pull these results together. The studies by Vining (1982, 1995) and Van Court and Bean (1985) were summarized in sections 5 and 6 of this chapter. To these

Table 6.11
Negative Correlations between Intelligence and Fertility in American Whites and Blacks

Study	Sex	Whites	Blacks
Vining (1995)	Females	- .06	- .23
Van Court and Bean (1985)	Combined	- .16	- .35
Broman et al. (1987)	Females	- .12	- .22
Average		- .11	- .27

are added the data reported by Broman and her colleagues at the National Institutes of Health (Broman, Nichols, Shaughnessy and Kennedy, 1987). Their study was based on approximately 17,000 white and 19,000 black babies born in the late 1970s. A great deal of information was collected at the time of their birth, and in subsequent years included the mothers' IQs, educational achievement and number of children. The IQs of the mothers correlated negatively with number of children at −0.22 for blacks and −0.12 for whites. The results of the three studies are summarized in Table 6.11, and the correlations averaged in the bottom row give negative correlations between intelligence and fertility of −0.11 for whites and −0.27 for blacks.

The greater dysgenic fertility among blacks than among whites has also been found by Herrnstein and Murray (1994) in their analysis of the data of the National Longitudinal Study of Youth. They do not report correlations between intelligence and fertility, but they calculate that white women who have children score at the 44th percentile of all white women, while among blacks the childbearers stand at the 42nd percentile and among Hispanics they stand at the 40th percentile. Thus, in all three groups childbearing women are less intelligent than the childless, and this trend is least pronounced among whites and most pronounced among Hispanics, with blacks falling intermediate between the other two groups. Notice that once again the childless have higher average IQs than those who have children.

The effect of greater dysgenic fertility among blacks and Hispanics as compared with whites must mean that the genetic quality of the black and Hispanic population has been deteriorating at a greater rate than that of the white. The conclusion that dysgenic fertility is greater among American blacks than among whites is corroborated by studies of the relation between educational level and fertility reviewed in Chapter 9 and between socioeconomic status and fertility reviewed in Chapter 10. The results hang together consistently. Genetic deterioration for American blacks, and probably also for Hispanics, is proceeding more rapidly than it is for whites.

10. DYSGENIC FERTILITY IN MALES AND FEMALES

Several of the studies reviewed in this chapter have found that dysgenic fertility for intelligence is considerably greater in females than males. This sex difference has also been found in a number of countries when educational attainment has been examined in relation to fertility for men and women separately, as we shall see in detail in Chapter 9.

Probably the explanation for this sex difference is that children impose a greater cost on the careers and other aspirations of intelligent and well-educated women than on those of intelligent and well-educated men, and also that women have a shorter period of childbearing years. It is women who have to bear most of the burden of childbearing and childrearing and who therefore have stronger incentives to limit their number of children.

At the other end of the intelligence spectrum, low IQ women tend to have higher fertility because they are inefficient users of contraception and there are always plenty of men willing to have sex with them. Low IQ men, on the other hand, tend not to have such high fertility because many of them are unattractive to females and lack the social and cognitive skills required to secure sexual partners.

A second factor accounting for the greater dysgenic fertility of women is probably their shorter span of childbearing years. Many intelligent women undergo prolonged education and devote themselves to their careers in their twenties and into their thirties, intending to postpone childbearing during the years when less intelligent women are having children. By the time childless, high-IQ, career women are in their thirties, significant numbers of them discover that they have waited too long to find suitable partners with whom to have children, or that they have become infertile. Older intelligent men who delay marriage and children until their thirties can find young wives more easily than older women can find young husbands, and they are less likely to become infertile.

11. CONCLUSIONS

In this chapter we have reviewed nine studies of the relationship between intelligence and fertility in the United States. The first four found that the relationship was positive, but they all had sampling deficiencies. The second set of five, carried out by Osborne, Vining, Van Court and Bean, Retherford and Sewell, and Herrnstein and Murray were based on better samples and all found negative relationships. This second set of studies, indicating the presence of dysgenic fertility, is to be preferred to the first, and, therefore, it is concluded that dysgenic fertility for intelligence has been present in the United States during the twentieth century.

The Van Court and Bean study shows that dysgenic fertility has been present for three generations from those born around 1890 to those born between 1960 and 1964. Dysgenic fertility was greater among the earlier birth cohorts born before 1909 than for later birth cohorts. This is consistent with the intelligence and sibling-size studies reviewed in Chapter 5, which also showed that dysgenic fertility slackened but did not disappear among cohorts born later in the twentieth century. Probably the reason for this is that knowledge and efficient practice of contraception began first in the closing decades of the nineteenth century among the more educated and intelligent classes, producing strong dysgenic effects, and then spread to the less educated and intelligent, mitigating the dysgenic trend.

The most satisfactory study for estimating the magnitude of the decline of genotypic intelligence is that of Retherford and Sewell, which indicates for whites born about 1940 a decline of 0.66 IQ points a generation. The decline must have been greater among earlier birth cohorts, because the negative association between intelligence and fertility was greater, as shown in the Van Court and Bean study. Probably the total decline for the three generations from the 1890s onward shown by Van Court and Bean has been around 2.5 IQ points. The magnitude of the decline appears to have been about two and a half times as great among blacks as whites, implying a decline among blacks of around 6.2 IQ points over the three generations. It seems probable that the continuing disadvantaged position of blacks in the United States in regard to educational attainment and employment is to some significant extent due to the greater deterioration of their genotypic intelligence.

Chapter 7

Intelligence and Fertility in Europe

1. England. 2. Scotland. 3. Greece. 4. Sweden. 5. Conclusions.

In the last chapter we examined the nine American studies on the relation between the intelligence of adults and their number of children and concluded that the weight of the evidence indicates that the association has been negative throughout the twentieth century. In this chapter we consider the evidence on the same question for Europe.

1. ENGLAND

The first study to be published in England on this issue was carried out by Young and Gibson (1965). Their sample consisted of 100 men and their wives in Cambridge born approximately over the years 1890–1900. The correlation between their IQs, presumably the average of the husband and wife, and number of children was 0.03 and should be considered as zero. Removal of the childless from the sample produced a residual correlation of −0.25, so the childless must have come predominantly from the low IQ bands.

There are several problems with the study. First, the sample size of 100 married couples is very small. Second, the authors are very economical about providing descriptive statistics: no information is given about the representativeness of the sample for socioeconomic status or intelligence, or the mean number of children. Third, the finding that the childless are of lower than average intelligence is contrary to a considerable amount of evidence reviewed in section 4 of Chapter 5 and confirmed by the Van Court and Bean (1985) study summarized in section 6 of Chapter 6, which shows that the childless are of higher than average intelligence. Fourth, the sample was certainly not representative of the general population because it consisted solely of married couples

Table 7.1
Intelligence and Fertility in England

	Nondeprived	Deprived	Multiply Deprived
Measure	Mean	Mean	Mean
IQ at age 11	104.3	93.9	86.7
Number of siblings	2.2	3.4	4.3
Number of children	1.9	2.3	2.7

Source: Kolvin et al. (1990).

and therefore excluded single mothers, who are known to be of below-average intelligence, and the unmarried, who are predominantly of above-average intelligence. The inclusion of these two groups would have swung the correlation negative. Hence, the best reading of the study is that if the sample had included the unmarried, fertility would have been dysgenic, in line with the better-sampled American study.

The second study of intelligence and fertility in England was carried out by Kolvin, Miller, Scott, Gatzanis and Fleeting (1990). The sample consists of 750 children born in the city of Newcastle in 1946. These children were divided into three groups according to the economic and social circumstances of their parents. The three groups were designated the Nondeprived (those with a comfortable living standard); the Deprived (those with a slightly below-average living standard); and the Multiply Deprived (the poorest group whose families were long-term unemployed). The children were tested for intelligence at the age of 11, and it was found that the mean IQs of the children in the three socioeconomic-status categories were 104.3, 93.9 and 86.7, respectively.

These children were followed up at the age of 33 and their average number of siblings and children recorded. The results are shown in Table 7.1. It will be seen that there are pronounced negative associations between the three groups and both their number of siblings and their number of children. The negative association is stronger for number of siblings—a measure of the strength of dysgenic fertility in the parents' generation born around 1920—than in the children's generation, born in 1946. This confirms the evidence of a slackening of dysgenic fertility in the middle decades of the twentieth century in England, which appeared in the reduction of the negative association between intelligence and number of siblings shown in Chapter 5. Notice, however, that among this cohort born in 1946 dysgenic fertility is pronounced, with the poorest group

Table 7.2
Average Number of Children of Men and Women in Scotland Broken Down by IQ Bands

	IQs of Men and Women								
	70-9	80-9	90-9	100-9	110-9	120-9	130-9	140-9	150-9
N. Men	8	43	57	50	46	39	12	12	1
N. Children	2.50	2.49	2.18	2.52	2.02	2.03	1.60	2.08	0
N. Women	11	49	74	51	33	18	14	4	2
N. Children	2.36	1.82	2.22	2.20	2.73	2.78	1.93	2.25	2.50

Source: Maxwell (1969).

having 32 percent more children than the Nondeprived group. This is what would be expected because there is considerable evidence, reviewed in Chapter 11, that the intelligence of children is positively related to the socioeconomic class of their parents.

2. SCOTLAND

In 1932 the intelligence of all eleven-year-old children in Scotland was tested, and a representative sample of 1,000 was selected from these and given the American Stanford-Binet Test. In 1968, an attempt was made by Maxwell (1969) to trace this "Binet Thousand," as it came to be known, and he succeeded in finding 517 for whom it was possible to obtain information on their number of children. At the time of this follow-up the sample was aged 47, so their fertility can be considered complete. From the published data, it is possible to calculate the number of men and women falling into nine IQ bands from 70–79 to 150–159 and the average number of children produced by these. The figures are shown in Table 7.2. Notice that for the men, the less intelligent clearly had more children than the more intelligent: the four lowest IQ bands had consistently more children than the five higher IQ bands. This trend is not present among the women for whom the correlation looks about zero. Maxwell did not give the correlations between his sample's IQs and their number of children.

Maxwell did not calculate either the selection differential or the generational change in genotypic intelligence. This can be done as follows. First, we assume that the mean IQ of the parents in each band is the average of the band, that is, parents in the IQ band 70–79 have a mean IQ of 74.5. Second, we assume that the average IQ of the children in each band is the same as the average of their parents. Third, we calculate the mean IQs of the men and women and arrive at figures of 105.46 and 101.84, respectively. Fourth, we calculate the mean IQs of the children of the men and women and arrive at figures of 104.10 and 102.29, respectively.

Thus, for the men the mean IQ of the children is lower than that of their fathers by 1.36 IQ points, while among the women the mean IQ of the children is higher than that of their mothers by 0.45 IQ points. Taking the men and women together, the mean IQ of the children is the average of these two figures, and is therefore lower than that of their parents by 0.91 IQ points. To obtain the decline of genotypic intelligence, this figure needs to be multiplied by the heritability, and if this is set at .82, as proposed in Chapter 5, the result is .75 IQ points. This figure compares closely with the decline in genotypic intelligence of .66 IQ points for American whites born in 1940 calculated from the Retherford and Sewell study described in section 7 of Chapter 6.

It is possible to criticize the representativeness of the sample in this study. The original Binet Thousand was probably representative, but only 517 of them were traced for the fertility study. Those who were traced were of higher average intelligence than those who were not traced. In the case of the men, the mean IQ of the original sample was 99.9 and of women, 98.5 (Scottish Council for Research in Education, 1949), whereas among those traced it was 105.46 for the men and 101.84 for the women. Evidently, the Maxwell sample has a somewhat restricted range as compared with the original sample. The effect of this would be to reduce the magnitude of the overall dysgenic fertility.

A curiosity of the results is that the men showed dysgenic fertility, whereas the women showed a slight trend toward eugenic fertility. This difference between the sexes is contrary to most of the evidence which finds greater dysgenic fertility among females, reviewed in section 10 of Chapter 6, and for fertility in relation to educational level, reviewed in Chapter 9. Probably this anomaly is a sampling quirk arising from the fact that the six intelligent women in the Scottish sample happened to have fourteen children between them.

3. GREECE

A study of intelligence and fertility in Greece was carried out around 1950 by Papavassiliou (1954). The sample consisted of 215 men for whom measures of intelligence, socioeconomic status and number of children were obtained. The sample was divided into four socioeconomic-status categories, and the results are shown in Table 7.3. It will be seen that the mean IQs decline with falling socioeconomic status, as would be expected. Average number of children are shown in the right-hand column and show small numbers of children among professional and managerial families and steadily increasing fertility with decreasing intelligence. It is evident that fertility was strongly dysgenic in Greece at this time.

Table 7.3
Intelligence and Fertility in Greece

Socioeconomic status	N	Mean IQ	N. Children
Professional, managerial	41	117.2	1.78
Lower white collar, skilled workers	80	100.9	2.66
Semiskilled	27	91.0	4.00
Unskilled	67	82.2	5.56

Source: Papavassiliou (1954).

4. SWEDEN

A study of the relationship between intelligence and fertility in Sweden has been made by Nystrom, Bygren and Vining (1991). They obtained a sample of 1,746 individuals living in the county of Stockholm but excluded the city itself. This appears to be suburban Stockholm, because 89 percent of the sample are described as living in densely populated areas. The data on intelligence and number of children were collected in 1970, and the sample was aged 26 to 65. The sample was biased toward the middle class and urban suburbia. The children recorded are described as "the children who live or have lived in the home for a long time," and were not necessarily the biological children of the sample. In the case of the women, it can be assumed that virtually all the children would have been their own biological children, but for men this is less certain. Sweden has had high rates of illegitimacy since the middle decades of the twentieth century. Large numbers of single women have children and subsequently live with and sometimes marry men who are not the fathers of the children. This raises doubts about the data for the relationship between intelligence and number of children among the men.

The results of the study are shown in Table 7.4, which breaks down the sample into four IQ bands and gives the average number of children for each band for the age groups 46–55, 56–65 and for the total sample. The results for the older groups represent completed fertility and are more satisfactory than the results for the total sample for which there must have been a significant element of uncompleted fertility, since they include 26 to 45-year-olds.

It will be noted that for men the relationship between intelligence and fertility is positive in both the older cohorts and more so for the 46 to 55-year-olds than for the 56 to 65-year-olds. Among women, the relationship is strongly negative among the 56 to 65-year-olds and neutral among the 46 to 55-year-olds.

The Swedish results indicate eugenic fertility among men and dysgenic fertility among the older Swedish women, replaced by neutral fertility among the 46 to 55-year-olds. There are, however, three problems with the data. First, as

Table 7.4
Average Number of Children among Parents by Age and IQ Class in Sweden in 1970

IQ Class	Age 46-55 N=465	Age 56-65 N=356	All ages (26-65) (N=1746)
Men			
94 and lower	1.4	1.6	1.5
95-109	1.6	2.0	1.7
110-124	1.8	1.8	1.7
125 and over	2.3	2.1	1.8
Women			
94 and lower	1.7	1.7	1.7
95-109	2.2	1.5	1.8
110-124	1.8	1.5	1.7
125 and over	2.1	1.0	1.8

Source: Nystrom, Bygren and Vining (1991).

noted, the children were not necessarily the biological children of the sample. Second, the geographical location of the sample, drawn from suburban Stockholm, raised the question of whether there is not a large rural population in Sweden which has lower than average intelligence and higher than average number of children. This is what has generally been found among rural populations. It has been documented for the United States by McNemar (1942) and there is further evidence from Britain and France for below-average intelligence among farm workers by Duff and Thomson (1923), Zazzo (1960) and Olivier and Devigne (1983), whose results are summarized in Table 11.2, and for high fertility among farm workers, summarized in Table 10.1. If this is the case in Sweden, and it probably is, the association between intelligence and fertility would be shifted toward the negative. Third, the study only recorded children living in the parental home, and a number of those aged 46 and older must have had adult children who were not living at home. These would have been disproportionately the less intelligent and the less educated, because these typically have children at younger ages. These children would have become adults and left home sooner, and their omission would have underestimated the fertility of the less intelligent. This would make the results appear less dysgenic than is actually the case. Taken together, these three problems with the sampling render the study inconclusive.

5. CONCLUSIONS

The evidence on intelligence and fertility in Europe is disappointingly meager. Nevertheless, the studies in England, Scotland and Greece show the same negative associations between intelligence and fertility that appear in major American studies reviewed in Chapter 6. The most satisfactory of the European investigations in respect of sample size and representativeness is the Scottish study. This showed a closely similar degree of dysgenic fertility as the leading American study of Retherford and Sewell, the two studies indicating a generational decline of genotypic intelligence of .75 and .66 IQ points, respectively. The Swedish study is challenging insofar as it appears to indicate eugenic fertility among men and among the younger cohort of women, but the sampling flaws are so serious that it has to be discounted.

Chapter 8

Resolving the Paradox of the Secular Rise of Intelligence

1. Increases in Intelligence in the United States. 2. Britain. 3. New Zealand. 4. Japan. 5. Continental Europe. 6. Two Interpretations of the Secular Increase of Intelligence. 7. Explaining the Secular Increase in Intelligence. 8. A Resolution of the Paradox. 9. Conclusions.

We have now seen that studies showing the inverse relationship between the intelligence of children and their number of siblings and between the intelligence of adults and their number of children both point to the same conclusion that the intelligence of the populations of the economically developed nations must be declining. However, this is only a predicted decline, and those who have been concerned with this question have naturally been interested in finding out whether this predicted decline in intelligence has actually been occurring. Over a period of approximately 70 years, from the 1920s to the 1990s, numerous studies have shown that the average intelligence level in a number of economically developed nations has, in fact, been rising over the course of most of the twentieth century. We are therefore confronted with the paradox of a predicted fall in the level of intelligence and an actual rise, which we shall attempt to resolve in this chapter. We begin by reviewing studies showing the secular rise of intelligence from the 1920s onward.

1. INCREASES IN INTELLIGENCE IN THE UNITED STATES

The first major study to attack the issue of whether intelligence has been declining was carried out by Smith (1942) on schoolchildren in Hawaii over the years 1924 to 1938. The result showed that the average level of intelligence had increased by thirteen IQ points over this fourteen-year period. This was a very large increase.

However, it did not particularly impress eugenicists in the 1940s and 1950s and was not cited by Burt (1946), Thomson (1949) or Cattell (1951) in their discussions of this question. Perhaps they did not know about the study, or perhaps Hawaii seemed to be something of a special case with its large number of Japanese, Chinese and other immigrants who were in the process of learning English and becoming acculturated in other ways. These unusual conditions might explain the rise in test scores and the apparent increase was evidently not considered to be persuasive.

The next study was published by Tuddenham (1948) and consisted of a comparison of the IQs of American military conscripts in the First and Second World Wars. The result showed that the average IQ of the conscripts had risen by thirteen IQ points over an interval of around 25 years. Here again, however, the result was not generally regarded as conclusive. As Vernon (1979) and others noted, conscripts in World War II had received considerably more education than those in World War I. Education may increase intelligence—or, at any rate, the cognitive skills assessed by intelligence tests—and this might be sufficient to explain the increase in the average IQs of military conscripts in the two World Wars.

For the next quarter of a century, there was a curious lack of interest in the United States in the question of whether the intelligence of the population was falling or rising.

It was not until the middle 1970s that R.L. Thorndike (1975) was to devise a new technique for tackling this issue. His method was to compare the IQs obtained by children and adolescents who had taken both the 1932 Stanford-Binet Test and the later version of the test standardized in 1972. Taking the results as a whole, he found that the subjects scored an average of 9.9 IQ points higher on the earlier test, indicating a rise in the intelligence of the population of this amount over the 40-year period. Closer examination of the data showed that most of this gain had occurred among preschool children. Thorndike inferred that some environmental factors must be having temporary boosting effects on the intelligence of young children, which largely washes out by the time they become adults.

A much more extensive analysis of the secular rise in intelligence in the United States using the same method was carried out some years later by James Flynn (1984). His method was to examine eighteen studies where various versions of the Wechsler and Stanford-Binet tests, standardized at different dates, had been given to the same sample of individuals. In general all these studies showed that people obtained higher IQs on the tests that had been standardized earlier. For instance, if in the year 1980 a person took a test which was standardized in the year 1950, he found the test easier and obtained a higher IQ than if he took a test standardized in 1970. The reason for this had to be that the population on which the earlier test was standardized had a lower average IQ than the population on which the later test was standardized. The difference between the IQs obtained on the two tests is a measure of the increase that has taken place in the intelligence of the population over the years between the standardizations of the two tests. Flynn used this method to measure the rise of intelligence in the United States over the 46-year period 1932–1978. His study yielded four important conclusions:

1. The average gain of the eighteen studies amounted to 13.8 IQ points over the 46-year period, representing a gain 0.3 IQ points per year or three IQ points per decade.
2. The rate of gain was constant over the 46-year period.
3. The rate of gain was the same among preschool children aged 2 to 6, among schoolchildren aged six to sixteen, and among adults. This was contrary to Thorndike's earlier conclusion derived from just one study and has to be accepted as a more reliable result. The interest of this result is twofold. First, it shows that the factors responsible for the rise in intelligence have been operating fully on 2 to 6-year-olds. This rules out any possible effect of improvements in education and focuses attention on effects operating early in life. Second, the fact that the gains have been as great among adults as among children and adolescents shows that they are not due to earlier maturation, leaving no effect on the level of intelligence finally achieved by adults.
4. The rise in intelligence was greater for the nonverbal and visuospatial abilities measured by the performance tests in the Wechsler Scales, where it amounted to around four IQ points per decade, than for the verbal tests, where it has been around two IQ points per decade. This tells against possible effects of improvements in cognitive stimulation, which would be expected to have greater effects on verbal tests measuring abilities like size of vocabulary and general information.

Flynn's studies of rising American IQs over the years 1932–1978 have been updated for the years 1972–1989 and 1978–1989 in Lynn and Pagliari (1994). The data are derived from administrations to the same group of subjects of the WISC-R and WISC-III, standardized in 1972 and 1989, respectively, and the WAIS-R and the WISC-III, standardized in 1978 and 1989, respectively. The results show that the earlier tests were still easier than the later ones. The mean IQ differences were 5.7 IQ points for the period 1972–1989 and 3.9 IQ points over the period 1978–1989. These represent secular IQ gains of 3.3 and 3.5 IQ points per decade; are broadly consistent with Flynn's estimates for the years 1932–1978; and show that the secular increases in intelligence were still taking place in the 1970s and 1980s. Once again, visuospatial tests showed greater rises than verbal.

2. BRITAIN

The years 1949–1950 saw the first of a number of studies showing secular increases in intelligence in Britain. The first of these came from Scotland. In 1932 the intelligence of all Scottish 11-year-olds, numbering 70,805, was assessed with a group verbal-educational test. The same test was administered again to all Scottish 11-year-olds in 1947. The 1947 children showed a gain of

2.2 IQ points (Scottish Council for Research in Education, 1949). The second study was carried out by Emmett (1950) and compared the scores of approximately 30,000 11-year-olds tested in England in 1938 and a further sample of approximately the same number in 1947. There was a very small increase over the ten-year period of .03 IQ points. The third study was carried out by Cattell (1951). Having tested the IQ of a large sample of 9 to 11-year-olds in the city of Leicester in 1936, Cattell retested the same-age cohort with the same test in 1949. He obtained an increase of 0.77 IQ points over the thirteen-year period.

These three British studies showing secular increases in intelligence were not open to the objection that increases in the amount of education could be responsible, since the samples tested were all at school and hence matched for the amount of education they had received. The results seemed to show that there must be something wrong with the theory that the intelligence of the population was in decline.

Some, like Duncan (1952) in the United States, thought that the results decisively demolished the theory that the national intelligence was in decline in the United States, Britain and other developed nations. Others, like Burt (1952), thought some artifact was involved, of which the most likely was an increase in test sophistication. Burt argued that in the 1930s, British children were unfamiliar with intelligence tests. By the late 1940s, virtually all British children were practiced on the tests in order to take the Eleven Plus Examination. This examination was introduced in Britain in 1944 to select 10-year-olds for different types of secondary school. On the basis of their performance, children with high IQs were allocated to grammar schools and the remainder to so-called secondary modern schools. Children were given practice for this examination, and it was soon discovered that practice increased the IQ obtained by several IQ points. Hence, Burt and others argued that practice on the tests was sufficient to explain the increase in intelligence obtained in the three studies and might even mask a real decline. Whether or not this explanation was correct, it began to become evident that it was not as easy to carry out a direct test of the theory that intelligence was in decline as had previously been thought.

Following these three studies there was little further interest in this question in Britain. However, it is possible to calculate secular increases in intelligence from a number of tests which have been standardized at one date and restandardized some years later. Where this has been done a comparison of the means of the two standardization samples provides a measure of the secular change in intelligence. A number of such studies have been assembled in Lynn and Hampson (1986) and are summarized in Table 8.1, together with the results of two more recent studies shown in the two bottom rows of the table.

All the studies show that secular increases in intelligence have been taking place over the period from 1932 up to 1986. The two most recent studies show increases over long time periods of 54 years in Scotland and 50 years in England. The recent Scottish study is based on a readministration in Scotland in 1986 of the test used in the Scottish surveys of 1932 and 1947 (Lynn, Hampson

Table 8.1
Ten Studies Showing Secular Increases of Intelligence in Britain

Time Period	Age (yr)	IQ Increase	Increase per decade	Test	Reference
1932-47	11	2.21	1.47	Verbal educational	Thomson, 1949
1938-47	11	0.03	0.03	Verbal reasoning	Emmett, 1950
1936-49	9-11	0.77	0.60	Non-verbal reasoning	Cattell, 1951
1943-79	11-15	2.70	0.75	Vocabulary	Lynn & Hampson, 1986
1945-57	11	6.12	4.71	Verbal reasoning	Lynn & Hampson, 1986
1959-77	9-11	3.43	1.90	Verbal reasoning	Lynn & Hampson, 1986
1938-79	8-14	7.63	1.86	Non-verbal reasoning	Lynn & Hampson, 1986
1949-82	5-11	8.79	2.66	Non-verbal reasoning	Lynn & Hampson, 1986
1954-77	7-8	8.47	3.68	Non-verbal reasoning	Lynn & Hampson, 1986
1955-74	11-12	0.94	0.41	Non-verbal reasoning	Lynn & Hampson, 1986
1935-85	10-11	12.50	2.50	Non-verbal reasoning	Lynn, Hampson & Mullineux, 1987
1932-86	11	5.85	1.08	Verbal-educational	Lynn, Hampson & Howden, 1988

Source: Lynn and Hampson (1986).

and Howden, 1988). This showed an increase of 5.85 IQ points over the 54-year period, representing a rise of 1.08 IQ points per decade. The relatively small increase reflects the verbal-educational nature of the test and is fairly typical of tests of this kind. The second recent study consisted of a readministration of Cattell's 1935 culture fair test to a sample of English children in 1985. The result showed a rise of 12.50 IQ points over the 50-year period, representing 2.5 IQ points per decade (Lynn, Hampson and Mullineux, 1987). Thus, all of the twelve sets of data show that there have been secular increases in intelligence in Britain of about the same magnitude as those which have been found in the United States.

3. NEW ZEALAND

The next country where secular increases in intelligence were observed was New Zealand. An American test, the Otis, was standardized in New Zealand in 1936 on 26,000 schoolchildren aged 10 to 13. The schools in which the testing was carried out were a representative sample of one-fifth of all schools in New Zealand. In 1968 the same test was administered to a new representative sample of 4,000 children and the results compared with the earlier sample by Elley (1969). Comparison of the scores of the two samples showed a 7.73 IQ point rise over the 32-year period, representing an increase of 2.42 IQ points per decade. The Otis test is largely verbal, consisting of vocabulary, verbal reasoning, comprehension, information and arithmetic. The rate of gain of intelligence in New Zealand was closely similar to that on verbal tests in the United States and Britain.

4. JAPAN

Four studies have been made of secular increases in intelligence in Japan covering the period 1953–1975. The studies and methods of calculating the IQ increases are fully described in Lynn and Hampson (1986) and a summary is given here. The results of the four studies are shown in Table 8.2. The average of the increases in mean IQ is 9.4 IQ points per decade. This is much greater than the increases that have taken place in the United States and Britain. The reason for this may lie in the privation suffered in Japan during and after World War II, which may have depressed intelligence levels, and the rapid postwar economic recovery.

5. CONTINENTAL EUROPE

By the middle 1980s, it was clear that intelligence had been increasing substantially in the United States, Britain, New Zealand and Japan over the course of the mid–twentieth century. Any lingering doubts about these increases were finally dispelled by Flynn's (1987) compilation of gains in intelligence in six

Table 8.2
Increases in Intelligence in Japan over the Period 1953–1975

Time Period	Age (yr)	IQ Increase	Increase per decade	Test	Reference
1953-60	9-15	10	14.3	General	Ushijima, 1961
1954-72	10-11	16.6	9.2	General	Sano, 1974
1951-75	6-16	13.8	5.7	WISC	Lynn & Hampson, 1986
1951-75	10	20.0	8.3	WISC	Lynn & Hampson, 1986

countries in Continental Europe. In all of these, considerable increases in intelligence were registered in the post–World War II decades. Some of Flynn's data were for gains on the Wechsler test made by young adolescents in Germany and Switzerland and these were broadly comparable to the gains in the United States, that is, about three IQ points per decade. Other results were derived from 18 and 19-year-old conscripts into the military in the Netherlands, Belgium, France and Norway and tested with Raven's Progressive Matrices. These showed rather larger secular increases. The difference between the two rates of gain is not attributable to age because adults show the same gains on the Wechsler tests as children (Flynn, 1984). Probably, therefore, the larger gains of the military conscripts is due to the nature of the test. The most probable explanation is that the Progressive Matrices consist largely of mathematical series problems in diagram format. The ability to solve these problems is almost certainly enhanced by studying math in school. In Europe adolescents have stayed in school longer in more recent post–World War II decades. With more schooling, they will have learned more mathematics, and this knowledge will have been applied to performance in the Progressive Matrices. These large gains in the Progressive Matrices, therefore, should probably be regarded as spurious. The same objection cannot be made to the Wechsler gains in Germany and Switzerland because the gains were made by younger adolescents at school throughout the period in which the gains were made—and the Wechsler test consists of a much wider range of problems, most of which are not taught in schools. The upshot of these results from Continental Europe is that the increases in intelligence have been of about the same size as those in the United States: around three IQ points per decade.

6. TWO INTERPRETATIONS OF THE SECULAR INCREASE OF INTELLIGENCE

Although it is indisputable that the average intelligence test scores of the populations of the economically developed nations have increased considerably from the 1930s to the middle 1980s, there is no agreed consensus on the interpretations of these rises. There are those who believe that the rises are spurious in the sense that they do not reflect real increases in intelligence, while others believe that the rises should be accepted at face value as genuine. The leading exponent of the view that the rises are artifactual is Flynn (1987). He asserts that it is not possible for intelligence to have increased by around fifteen IQ points from the 1930s to the 1980s, because if it had done so, it would be immediately apparent that people today are much more intelligent than their grandparents of 50 years ago. College professors, in particular, would have noticed that their students had become much cleverer in recent years, but there has been no general recognition that this is the case.

My own view differs from that of Flynn and regards the increases in intelligence as measured by the tests as genuine. There are four principal reasons for taking this view.

1. The studies of the secular rise in intelligence show that the increase has been relatively small among the higher IQ groups as compared with the lower IQ groups. This has been found in both Britain and Denmark (Lynn and Hampson, 1986; Teasdale and Owen, 1989). Many of those who have not noticed a secular rise of intelligence are those who have high IQs themselves and had parents and grandparents with high IQs, so they are not likely to be conscious of a secular rise in their own families.
2. The average height of the population increased from the 1930s to the 1990s by about the same amount, namely, one standard deviation, as the increase in intelligence. Yet it is doubtful whether many people are conscious in the 1980s and 1990s that they are significantly taller than their grandparents. Similarly, there have been increases in running speeds, jumping heights, and so forth, as indexed by improving performance in the records achieved in Olympic games; but again it is doubtful whether many people in the 1980s and 1990s are conscious that they can run faster and jump higher than their grandparents.
3. Intelligence tests are too well validated against real-life success in educational and occupational achievement for the increase registered in all abilities over the course of the middle decades of the twentieth century to be artifactual.
4. The fact that the secular increases in intelligence have been as great among preschool children as among adolescents and adults argues against alternative explanations in terms of improvements in test sophistication and greater familiarity with the content of the tests, which would be expected to have minimal effects on preschool children.

For these reasons I believe that the increases in intelligence have to be accepted as genuine, a view with which Brody (1992) concurs.

7. EXPLAINING THE SECULAR INCREASE IN INTELLIGENCE

Two principal theories have been advanced to explain the secular increase in intelligence, which has occurred in economically developed nations during the twentieth century. The first attributes these to increase in cognitive stimulation resulting from improved education; the greater availability of cognitively stimulating books, toys and television; and the reduction in family size which has enabled parents to give their children more attention. The second theory is that increases in intelligence have been due to improvements in nutrition and related physiological factors, such as the reduction in the debilitating effects of infectious diseases. It is possible to hold, like Brody (1992) and Storfer (1990), that both factors have contributed to the secular increase in intelligence.

The most plausible of these theories is that the major factor has been improvements in the quality of nutrition received by the fetus and by young babies. I have set out the detailed arguments for this conclusion in Lynn (1990) and summarize here the principal arguments for this view.

1. There is direct evidence that the quality of nutrition received by the fetus has a permanent effect on subsequent intelligence. This comes from studies of identical twins who are occasionally born with different birth weights as a result of one twin receiving better nutrition from its mother's placenta than the other. In these cases the heavier twin at birth has greater intelligence in adolescence. Six studies of this kind have been published and all report the effect (Churchill, 1965; Kaelber and Pugh, 1969; Scarr, 1969; Babson and Phillips, 1973; Fujikura and Froehlich, 1974; Hendrichsen, Skinhoj and Andersen, 1986). These studies demonstrate that the quality of prenatal nutrition has a permanent effect on intelligence.
2. Several of these studies show that the higher intelligence of the heavier twin at birth is greater for nonverbal and visuospatial abilities than for verbal abilities. This is consistent with the greater secular increase in the nonverbal and visuospatial abilities as the quality of nutrition has improved over the course of the twentieth century. The greater rate of increase of the nonverbal abilities is difficult to explain by the cognitive stimulation theory, from which a greater increase in verbal abilities would be expected.
3. Real standards of living in the economically developed nations approximately doubled over the half century from the 1930s to the 1980s (Coleman and Salt, 1992). As a result, the populations have been able to afford higher quality nutrition. The effect of this has been increases in the birth weights of babies (Gruenwald, Funakawa, Mitani, Nishimura and Takeuchi, 1967) and increases in height and brain size (Miller and Corsellis, 1977). Numerous studies have shown that brain size is positively associated with intelligence: eleven such studies are summarized in Lynn (1990). The secular increase in height and brain size from the 1930s to the 1980s has been of about the same order of magnitude as the increase in intelligence, namely, one standard deviation (Whitehead and Paul, 1988).
4. Alternative theories that the secular increase in intelligence is due to improvements in cognitive stimulation resulting from better education, greater availability of books, toys, TV, and so on do not stand up to scrutiny. Any possible effect of improvements in education can be ruled out by the fact that the full magnitude of the increase in intelligence has occurred among 4 to 6-year-olds (Flynn, 1984) and is even present in the improved development quotients of 2 and 3-year-olds (Lynn, 1990).
5. It is doubtful whether cognitive stimulation in the family has any permanent effect on intelligence which lasts into adulthood. The reason for this conclusion lies in the zero correlations between the IQs of pairs of unrelated adopted children reared in the same family (Scarr and Weinberg, 1978; Teasdale and Owen, 1984). If cognitive stimulation had an effect it would produce positive correlations because families differ in the quantity and quality of the cognitive stimulation they provide. The fact that the correlation is zero indicates that differences in cognitive stimulation have no permanent effect, although they may have temporary effects on the intelligence of children which fade out when they become adults.

For these reasons I believe there is no particular mystery about the increase which occurred in the intelligence of the populations of the economically developed nations over the middle decades of the twentieth century. The increase in living standards brought about an improvement in nutrition. This produced improvements in physical development, including heavier babies with larger brains and probably improved neurological development, which in turn have brought about greater intelligence.

8. A RESOLUTION OF THE PARADOX

There is an abundance of evidence that in the economically developed nations during the twentieth century, people with higher intelligence had fewer children than those with lower intelligence. The effect should have been that the level of the intelligence of the populations declined in the middle decades of the century. This effect was confidently predicted by people like Cattell (1937), Burt (1946) and Thomson (1949). Yet the predicted decline did not occur. On the contrary, the level of intelligence of the populations of the economically developed nations increased substantially over the course of the century. How can this paradox be resolved? Three solutions have been offered for this problem. The first was advanced by Higgins, Reed and Reed (1962) and Bajema (1963). They proposed that the supposed negative correlation between intelligence and fertility did not exist and that the true correlation has been slightly positive. If this is so, the intelligence of the population would remain approximately stable or increase slightly. There are two errors in this theory. First, better quality evidence collected by Vining (1982), Van Court and Bean (1985) and Retherford and Sewell (1988) in the United States and Maxwell (1969) in Scotland have established that the association between intelligence and fertility is indeed negative, as the earlier investigators inferred from the negative associations between intelligence and number of siblings. Second, even if the association between intelligence and fertility were positive, the magnitude of the association claimed by Higgins, Reed and Reed and by Bajema was much too small to account for the large increases that have taken place in intelligence, so the causes of this increase remain an unsolved problem.

A second solution to the paradox put forward by people like Duncan (1952) in the United States and Penrose (1950) in Britain was that there must be some flaw in the reasoning that led to the inference that a decline must be taking place. Penrose constructed a simple genetic model to explain how the intelligence of the population would remain stable if there were high fertility in the middle IQ bands, low fertility at both extremes and an overall negative association. The model does not, however, match the evidence which indicates a linear decline in fertility with rising intelligence, and the model does not explain the increase in intelligence levels.

The third solution was put forward by Burt (1946) who argued that there must be some environmental improvement which has masked the decline. He proposed that this environmental improvement was probably an increase in test sophistication which has led to spurious increases in test scores. This suggestion is almost certainly wrong, because the increases have taken place among 4 to 6-year-olds who have not experienced greater familiarity with the tests, and have been greater in the nonverbal abilities, which are probably less practiced. Nevertheless, Burt was essentially right in proposing that environmental improvements have masked the decline. These environmental improvements, however, have consisted largely of better nutrition and not of greater test sophistication.

In the 1940s and 1950s, it was supposed by people like Thomson (1949) and Cattell (1951) that it was possible to carry out a simple experiment to determine whether the intelligence of the population was falling. In a sense this was right, but in a more profound sense it was wrong. The ambiguity lies in the distinction between genotypic intelligence (the quality of the genes) and phenotypic intelligence (the scores obtained on the tests). This distinction was not sufficiently thought through by the people who were concerned with this problem in the middle decades of the century. It was easy enough to show that phenotypic intelligence has risen over the course of the twentieth century in many economically developed countries, but this does not really tell us anything about whether genotypic intelligence has risen, fallen or remained constant.

It is quite possible for the phenotypic intelligence of the population to show an increase while the genotypic quality is in decline. To take an agricultural parallel, it is possible to sow seeds of deteriorating genetic quality in successive years and to pump in so much more fertilizer that the size of the crop actually improves. This is probably what has happened with intelligence, although there is no way of making a direct test of the theory that the genotypic intelligence of the population is in decline. The theory is only an inference from the negative associations between intelligence and fertility, which the preponderance of the evidence indicates are present in the United States and Britain, buttressed by the negative associations between intelligence and number of siblings.

The best reading of the evidence is that genotypic intelligence (the genetic quality of the population) has been in decline in the economically developed nations over the course of the twentieth century, and probably since the birth cohorts of around 1830, when socioeconomic-status differences in fertility began to appear. This genotypic decline has been masked by environmental improvements consisting principally of better nutrition, which have brought about the phenotypic increases. If this is the correct view, the intelligence of the populations cannot be expected to continue rising indefinitely. On the contrary, diminishing returns to the improvements in nutrition will occur as the quality of nutrition reaches its optimum. When this occurs, and if the negative association between intelligence and fertility persists, phenotypic intelligence will start to decline.

9. CONCLUSIONS

From the closing decades of the nineteenth century through the 1940s, eugenicists believed that in Western nations intelligent people were having fewer children than the unintelligent, and therefore that the intelligence level of the populations must be deteriorating. But from the 1940s onward, studies began to appear showing that intelligence has been increasing, and by the 1990s this increase has been firmly established in many countries. There are differences of opinion about whether the increase is real or an artifact, perhaps due to greater familiarity with the tests. For various reasons, it is believed that the increase is wholly or largely genuine.

The increase in intelligence was a paradox which took some time to understand. Some thought there must be some flaw in the argument that the inverse association between intelligence and fertility would lead to an intelligence decline. Others continued to think there was a decline but that it was masked by increases in test-taking know-how. It is concluded that the cause of the increase in intelligence has been due principally to improvements in nutrition, which have brought about larger brain size and probably enhanced the neurological development of the brain. The solution to the paradox, therefore, lies in drawing a sharper distinction between genotypic intelligence, which is declining because of the inverse association between intelligence and fertility, and phenotypic (measured) intelligence, which has been rising because of environmental improvements and has more than counteracted the genotypic deterioration. These environmental improvements are bound to be subject to diminishing returns. When their impact is exhausted, and if dysgenic fertility continues, phenotypic intelligence will begin to decline.

Chapter 9

Education and Fertility

1. United States. 2. Britain. 3. Australia and Canada. 4. Europe. 5. The Netherlands. 6. Japan. 7. Singapore. 8. Heritability of Educational Attainment. 9. Conclusions.

There have been many more studies on the relationship between fertility and educational attainment than between fertility and intelligence. Educational attainment is quite closely related to intelligence. Numerous studies have shown that the two are correlated at a magnitude of around .6 (Eysenck, 1979; Brody, 1992). If fertility in relation to intelligence has been dysgenic in Western nations throughout the twentieth century, as we concluded in Chapters 4 through 6, we should expect this to show up in a parallel inverse relationship between educational attainment and fertility. A consideration of the evidence on whether this is the case forms the subject of the present chapter.

1. UNITED STATES

Data on the relationship between women's level of education and their number of children was collected in the United States censuses of 1940 and 1960, and in a survey carried out by the Bureau of the Census in 1990. The results have been analyzed by Osborn (1951), Kiser and Frank (1967) and Bachu (1991) and are summarized in Table 9.1. The first six rows of data are for completed or virtually completed fertility of women aged 35 and over. The last three rows give the anticipated final fertility of 18 to 34-year-old women, born between 1956 and 1972. The 1940 census returns were analyzed for white women only, the 1960 data for whites and blacks and the 1990 data for whites, blacks and Hispanics. The earlier studies divide educational attainment into four levels running from less than four years of high school to college degree, while the later

Table 9.1
Fertility of American Women in Relation to Educational Level

Sample	Date of Birth	Age	Educational Level					Dysgenic Ratio	Reference
			1	2	3	4	5		
White	1890-95	45-49	3.43	1.75	1.71	1.23	---	2.79	Osborn, 1951
White	1915-19	40-44	2.83	2.44	2.15	1.97	---	1.44	Kiser and Frank, 1967
Black	1915-19	40-44	3.20	2.69	2.12	1.66	---	1.93	Kiser and Frank, 1967
White	1946-55	35-44	2.57	2.01	1.72	1.86	1.59	1.49	Bachu, 1991
Black	1946-55	35-44	2.97	2.28	1.86	1.97	1.70	1.62	Bachu, 1991
Hispanic	1946-55	35-44	3.07	2.29	2.17	2.47	1.72	2.09	Bachu, 1991
White	1956-72	18-34	2.38	2.05	2.05	2.02	2.01	1.18	Bachu, 1991
Black	1956-72	18-34	2.63	2.23	2.00	2.01	1.91	1.34	Bachu, 1991
Hispanic	1956-72	18-34	2.74	2.21	2.14	2.18	2.01	1.31	Bachu, 1991

Note: Educational level: 1 = less than 4 years high school; 2 = 4+ years high school; 3 = some college; 4 = 3 years college; 5 = 4+ years college.

study adds a fifth educational level by subdividing those with tertiary education into basic college graduates and those with four and more years of university education. The data are for all women, not married women only.

If we look first at the general trends, we see that in all the samples the more poorly educated women had more children than the better educated. To provide an approximate measure of the magnitude of the dysgenic fertility, dysgenic ratios have been calculated by dividing the fertility of the least educated with that of the most educated. Because in the first three cohorts all college-educated women are aggregated into one category, while in the later cohorts this group is subdivided into three years of college and over four years of college, the last two categories have been combined for the purpose of calculating the dysgenic fertility ratios. This makes these ratios comparable across all the cohorts.

Looking at the secular trends in fertility in relation to educational level, we observe among white women dysgenic fertility in the first cohort born 1890–1895, among whom the dysgenic fertility ratio is 2.79, and slackening of dysgenic fertility among those born twenty years later to a dysgenic fertility ratio of 1.42. For the next cohort, born 1946–1955, the dysgenic fertility remains about the same. The final data for white women consist of the anticipated final fertility of the 1956–1972 cohort, and among these dysgenic fertility has slackened to 1.18. The anticipated fertility figures may not be realized because some women may have fewer children than they anticipate, perhaps because of infertility, or for various reasons, while others may have more, perhaps because of unplanned births. Nevertheless, they are the best we have for recent birth cohorts and will probably turn out to be approximately accurate.

Looking next at the blacks, we observe that these show the same initially high dysgenic fertility ratio among the 1915–1919 birth cohort of 1.93, slackening to 1.62 among those born 1946–1955, and then to 1.34 for anticipated final fertility among the most recent cohort born 1956–1972. The same secular trend is present among the two Hispanic cohorts, among whom dysgenic fertility declines from 2.09 in the 1946–1955 cohort to 1.31 among the 1956–1972 cohort. Thus, the general secular trend for dysgenic fertility to slacken, although not to be eliminated, is present for all three racial groups.

Turning now to the race differences in fertility, we can observe that the number of children of white women is lower than that of blacks at all educational levels, except for the college-educated 1915–1919 cohort; the Hispanics have the greatest number of children. Since numerous studies carried out since World War I have shown that whites have higher average intelligence levels than blacks and Hispanics by approximately fifteen and eight IQ points respectively (Herrnstein and Murray, 1994), this means that dysgenic fertility in the American population considered as whole is greater than the dysgenic fertility within the racial and ethnic subpopulations.

We look finally at the racial differences in the dysgenic fertility ratios and note that these are greater for blacks than for whites in all three cohorts for which data are available for both groups. This is consistent with the greater dysgenic fertility among blacks for intelligence found by Vining (1982, 1995) and by Van Court and Bean (1985) and reviewed in Chapter 6. The Hispanic dysgenic fertility ratio is the largest among the 1946–1955 cohort, but is slightly lower than the black among the 1956–1972 cohort.

2. BRITAIN

British data for fertility in relation to the educational level of women is collected for representative samples of the population in the General Household Survey (1989), an annual Government survey of some 20,000 adults in which information is collected on a variety of subjects. The results for completed fertility in relation to four educational levels are shown for the two birth cohorts

Table 9.2
Fertility of the 1918–1943 and 1935–1944 Birth Cohorts of British Women, and Anticipated Fertility of the 1960–1969 Cohort, in Relation to Educational Levels

Date of Birth	Age	N	1	2	3	4	Dysgenic ratio	Reference
1918-43	45-70	2,377	2.44	2.10	2.12	1.97	1.23	General Household Survey, 1989
1935-44	45-50	2,696	2.50	2.20	2.09	2.08	1.20	General Household Survey, 1993
1960-69	20-30	3,326	2.35	2.20	2.14	2.07	1.14	General Household Survey, 1993

Note: Educational level: 1 = no educational qualifications; 2 = minor qualifications; 3 = some GCSE grades A–C; 4 = some GCE A levels.
Source: General Household Survey (1989, 1993).

of 1918–1943 and 1935–1944 in Table 9.2; both show dysgenic fertility. The third row gives the anticipated number of children of 20 to 30-year-old women. The trend remains dysgenic but slightly less so than for the completed fertility of the older cohorts.

3. AUSTRALIA AND CANADA

The two major English-speaking nations apart from the United States and Britain are Australia and Canada. Data on the relation between the education of married women and their average number of children are shown in Table 9.3. The Australian figures and the first and third of the Canadian figures come from the censuses. The second set of Canadian data come from a survey of 2,082 35 to 49-year-old married women reported by Grindstaff, Balakrishnan and Dewit (1991). Note that the four data sets show dysgenic fertility in relation to women's levels of education, and that the magnitude of the dysgenic fertility ratio in Canada declines among the younger birth cohorts, as in the United States.

Table 9.3
Fertility of Married Women in Relation to Educational Level in Australia and Canada

Country	Date of Birth	Age	Educational Level				Dysgenic Ratio	Reference
			1	2	3	4		
Australia	1942-46	35-39	2.57	2.42	2.14	1.68	1.53	Rowland, 1989
Canada	1900-25	45-70	3.86	---	2.75	2.44	1.58	Census of Canada, 1971
Canada	1935-50	35-49	3.26	2.74	2.25	2.00	1.63	Grindstaff et al., 1991
Canada	1940-44	40-44	3.36	2.89	2.59	2.19	1.53	Census of Canada, 1981

Note: Educational level: 1 = basic; 2–3 = intermediate; 4 = tertiary.

4. EUROPE

Most of the data on the relationship between educational level and fertility are, for reasons best known to the demographers and sociologists who collect these figures, confined to married women. What about men? A useful study giving data on fertility in relation to educational level for men, as well as women, for eight countries of Continental Europe and England for the years 1966–1977 has been made by Nohara-Atoh (1980). His results are shown in Table 9.4. The figures are estimates of completed fertility of married women and men standardized by duration of marriage. There are two main features of interest in the data. First, in all the countries the better educated tend to have fewer children. The magnitude of the inverse relationship between education and fertility is indexed by the dysgenic fertility ratios which are all greater than one, indicating the percentage of greater fertility among the least educated in relation to that in the best educated. The second feature of interest in the data is that in all the countries the dysgenic fertility ratios are greater for women than for men. This confirms the results we noted for intelligence in Chapter 6. These dysgenic fertility ratios almost certainly underestimate the greater dysgenic fertility among women because they are derived from married women, and there is a stronger tendency for well-educated women to be unmarried than there is for unmarried men.

Table 9.4
Number of Children Ever Born to Married Women and Men Analyzed by Education, Standardized for Duration of Marriage

	Country	Date	Primary	Lower Secondary	Higher Secondary	Tertiary	Dysgenic Ratio
Women							
	Belgium	1966	2.60	2.00	1.95	2.07	1.26
	Czechoslovakia	1970	2.27	1.64	1.64	1.64	1.38
	Denmark	1970	2.12	1.80	1.83	1.89	1.12
	England	1967	---	1.86	1.73	1.69	1.10
	Finland	1971	2.43	1.60	1.60	1.86	1.31
	France	1972	3.26	1.92	1.92	1.89	1.72
	Hungary	1966	2.71	1.72	1.46	1.34	2.02
	Poland	1972	2.87	2.33	1.82	1.60	1.79
	Yugoslavia	1970	2.40	1.82	1.43	1.34	1.79
Men							
	Belgium	1966	2.40	1.94	2.17	2.07	1.16
	Czechoslovakia	1970	2.21	1.71	1.71	1.64	1.35
	Denmark	1970	2.11	1.85	1.87	1.79	1.18
	England	1967	---	1.85	1.70	1.72	1.07
	Finland	1971	2.20	1.92	1.92	1.80	1.22
	France	1972	3.09	1.97	1.92	2.08	1.48
	Hungary	1966	2.62	1.75	1.43	1.50	1.74
	Poland	1972	2.83	2.40	1.94	1.60	1.77
	Yugoslavia	1970	2.68	2.01	1.69	1.55	1.73

Source: Nohara-Atoh (1980).

5. THE NETHERLANDS

One of the problems in determining whether fertility is dysgenic is that people do not invariably complete their number of children until they reach their mid-forties in the case of women, or sometimes even later, in the case of men. This means that it is only possible to calculate whether fertility is dysgenic for cohorts born 45 years ago and not for those born more recently. One way of partially overcoming this problem is to ask younger people about how many children they expect to have. Although it is likely that these expectations are not always accurately fulfilled for a variety of reasons, they are probably roughly correct. In the first two sections of this chapter, anticipated fertility in relation to educational level has been given for the United States and Britain.

Table 9.5
Expected Number of Children and Expected Childlessness in The Netherlands

	Birth Cohort	Measure	Educational Level Low	Medium	High
Women	1950-59	Expected children - number	2.10	2.00	1.80
		Expected childless - percent	10.00	11.00	24.00
Women	1960-69	Expected children - number	2.20	2.20	1.90
		Expected childless - percent	8.00	10.00	16.00
Men	1950-59	Expected children - number	1.90	2.00	1.90

Source: Statistics Netherlands (1993).

Some useful data on anticipated number of children in relation to educational level were collected in The Netherlands in 1993 as part of a government social survey (Statistics Netherlands, 1993). Representative samples of several thousand men and women born between 1950 and 1959 and between 1960 and 1969, therefore aged 34 to 43 and 24 to 33 at the time of the survey, were asked how many children they expected to have and whether they expected to be childless, and the results were analyzed by three educational levels. Fuller information was collected for women than for men, and the results are shown in Table 9.5.

Notice that for women born 1950–1959 expected fertility is dysgenic with the least educated women expecting to have the most children. For men, expected fertility in respect of educational level is neutral. This difference is consistent with a number of other results we have seen showing that fertility is more dysgenic for women than for men. The anticipated dysgenic fertility for women is further illustrated by the percentages who expect to be childless, which are more than twice as great among the college-educated women as among those without a college education. This confirms yet again the conclusion that the childless tend to be the most educated.

The birth cohort of 1950–1959 has the same anticipated dysgenic fertility as the younger birth cohort of 1960–1969, but the older cohort has a substantially larger proportion of expected childlessness of 24 percent as compared with 16 percent among the younger cohort. Probably the anticipated childlessness of the 34 to 43-year-olds is fairly accurate because if a woman has not had a child by the age of 34 and expects to remain childless she will probably be childless. The anticipated childlessness of the younger women aged 24 to 33 is less likely to be accurate, and it is doubtful whether any significance can be attached to the differences between the two birth cohorts. The upshot of the study is that the fertility of the two cohorts of Dutch women is likely to be dysgenic, while that of Dutch men is likely to be neutral.

6. JAPAN

There are quite extensive data on the relationship between education and fertility in Japan. Ogawa and Retherford (1993) give the figures for married women aged 40 to 49 from surveys carried out over the years 1963–1992. The women were divided into the three groups of those with primary, secondary and teriary education, and their average number of children was calculated. The results are shown in Table 9.6. Dysgenic fertility ratios are given in the right-hand column. Notice that in the period 1963–1975 dysgenic fertility was around 1.30 and about the same as that in the United States, Canada, Britain and Europe generally. However, from 1977 onward dysgenic fertility weakened, and fertility even turned eugenic in the years from 1984 to 1988. The small annual fluctuations between eugenic and dysgenic fertility over the period 1981–1992 should probably be regarded as sampling errors, and the relation between education and fertility over this decade is best expressed as the average of the six results. This is precisely 1.00, indicating that fertility was neither eugenic nor dysgenic in this decade. This is an encouraging result for all who feel concern about dysgenic fertility, but unhappily the surveys are based on married women only, and well-educated women in Japan are less likely to be married than the poorly educated. Ogawa and Retherford estimate the percentages of women who will marry in Japan by the age of 40 are 95.6 percent for the least educated, 94.0 percent for those with intermediate-level education, and 86.2 for those with tertiary education. Adjusting the figures in Table 9.6 for these proportions of women who remain single and for the most part childless (children born out of wedlock are a further problem but are negligible in Japan) gives fertility rates for the three groups of women for 1992 as 2.16 (least educated), 2.00 (intermediate education) and 1.87 (college education) and a dysgenic fertility ratio of 1.16. Further evidence for dysgenic fertility in Japan, showing inverse relationships between the number of children of married women in relation to their husbands' educational level, has been published by Watanabe (1990) for the birth cohorts of 1885–1895 through 1935–1945.

7. SINGAPORE

Useful data on women's fertility in relation to their education and broken down by ethnic group are available in the Singapore census of 1980. It is shown in Table 9.7 for women aged 35 to 39, born between the years 1941 and 1946. The first three rows give the average number of children of Chinese, Malay and Indian married women broken down by five educational levels. Notice that in all three groups fertility is quite strongly dysgenic, with college-educated Chi-

Table 9.6

Fertility (mean number of children ever born) of Currently Married Women Aged 40–49 by Urban and Rural Residence and by Level of Education: Japan, 1963–1992

		EDUCATION			
Year of survey	Total	Primary	Secondary	Tertiary	Dysgenic Ratio
1963	3.21	3.37	2.94	2.64	1.28
1965	3.04	3.22	2.68	2.30	1.40
1971	2.51	2.68	2.27	2.17	1.24
1975	2.30	2.42	2.18	1.84	1.32
1977	2.24	2.41	2.08	2.11	1.14
1979	2.23	2.29	2.20	2.03	1.13
1981	2.23	2.23	2.23	2.04	1.09
1984	2.27	2.28	2.22	2.31	0.99
1986	2.20	2.22	2.04	2.28	0.97
1988	2.22	2.22	2.13	2.48	0.90
1990	2.14	2.16	2.07	2.09	1.03
1992	2.23	2.26	2.13	2.17	1.04

Source: Ogawa and Retherford (1993).

Table 9.7
Fertility of Women in Singapore Analyzed by Ethnic Group and Education

		Educational Level					
Group	Status	Basic	Primary	Secondary	Higher Secondary	Tertiary	Dysgenic Ratio
Chinese	Married	3.47	2.75	2.12	2.17	1.93	1.80
Malay	Married	4.33	3.89	2.69	2.85	0.70	6.18
Indian	Married	4.18	3.20	2.20	2.46	2.36	1.77
Total	Married	3.63	2.91	2.15	2.21	1.91	1.90
Total	All	3.50	2.73	1.93	2.02	1.65	2.12

Source: Singapore Census (1980).

nese and Indian women having barely more than half the children of women with less than full primary school education (''basic''), while among the Malay women dysgenic fertility is even greater. There are also ethnic differences in fertility such that the Chinese have the fewest children (overall average = 3.03), followed by the Indian (overall average = 3.41) and the Malay (overall average = 3.41). The interest of these ethnic differences is that the Chinese score on average about fourteen IQ points higher than the Malays, the mean IQs being 110 and 96, respectively (Lynn, 1977). Thus, the ethnic group differences in fertility and intelligence in Singapore reflect those in the United States—where the less intelligent group has higher fertility—and reinforces the dysgenic fertility of the population as a whole. It is interesting to note the very low fertility of Malay college graduates, which also mirrors the low fertility of black college graduates in the United States.

A further point of interest in the Singapore fertility data lies in the fertility of all women as compared with that of married women, shown in the last two rows of the table. Observe that the fertility of all women is more dysgenic than the fertility of married women, except for the Malays. The reason for this is that unmarried and normally childless women tend to be more concentrated among the better-educated. Because more better-educated women are unmarried, this underestimates the true magnitude of the dysgenic fertility. The Singapore data illustrate this point and show that the underestimate is 12 percent.

8. HERITABILITY OF EDUCATIONAL ATTAINMENT

We have now seen that throughout the economically developed world well-educated women have been having fewer children than poorly educated women

Table 9.8
Studies of the Heritability of Educational Attainment

Country	Date of Birth	Sex	Heritability	Reference
Australia	1893-1950	Males	.53	Baker et al.,1994
Australia	1893-1950	Females	.52	Baker et al., 1994
Australia	1951-65	Males	.56	Baker et al., 1994
Australia	1951-65	Females	.57	Baker et al., 1994
Norway	1900-35	Males	.10	Tambs et al., 1989
Norway	1935-	Males	.51	Tambs et al., 1989
Norway	1900-40	Males	.41	Heath et al., 1985
Norway	1900-40	Females	.41	Heath et al., 1985
Norway	1940-	Males	.74	Heath et al., 1985
Norway	1940-	Females	.45	Heath et al., 1985
Sweden	1935	Males	.48	Husen, 1959
United States	1900-45	Males	.56	Vogler and Fulker, 1983
United States	1917-27	Males	.44	Taubman, 1976
United States	1977-87	Males	.34	Petrill and Thompson, 1994
United States	1977-87	Females	.68	Petrill and Thompson, 1994
United States	1950	Males	.44	Loehlin and Nichols, 1976
United States	1950	Females	.41	Loehlin and Nichols, 1976

in the closing decades of the nineteenth century and in the twentieth. The next question is whether this is a genetic problem. It can be argued that differences in educational attainment are solely caused by the environmental advantages and disadvantages of being born into middle- or lower-class families and do not reflect genetic differences at all. If this were so, fertility differentials in relation to educational level would not be a matter of genetic concern.

The crucial issue is whether educational attainment has a heritability. The principal method for investigating this question has been by the comparison of the correlations for educational attainment of identical and nonidentical twins. The heritability is calculated either by doubling the difference or by more sophisticated methods of structural equation modelling. The results of seventeen studies are shown in Table 9.8. In a number of cases the heritabilities are averages of those obtained for several academic subjects. For instance, the Loehlin and Nichols (1976) results are the averages of the heritabilities of English, Math, Social Science, Science and Vocabulary. The data are derived from adults except for the Petrill and Thompson (1994) study which is based on 7 to 13-year-olds. Precise or approximate dates of birth of the subjects are given to allow examination of possible secular trends in heritability. The Tambs, Sundet, Magnus and Berg (1989) study indicates a considerable increase in heritability from .10 to .51 for men born before 1935 and after 1935, but this finding does not seem to be confirmed by other results, such as those from Australia. Results are also given for males and females, but there does not seem to be any significant sex difference in the heritability coefficients.

It will be noted that all seventeen studies show some heritability for various measures of educational attainment and all are appreciable with the exception of the .10 figure obtained by Tambs et al. for Norwegian men born early in the century. The average of the seventeen studies is .51. To this can be added a further study carried out in Denmark on pairs of adopted brothers born in the decade 1938–1947 from which Teasdale and Owen (1984) estimate a heritability of .68. The addition of this study gives an average of .52. This figure needs an upward adjustment because the measures of educational attainment are not wholly accurate, reliable or stable over time. It is not certain what the appropriate adjustment should be, so the figure of .52 for the heritability of educational attainment in Western nations will be allowed to stand. The point is that the figure is appreciable and means that the universally present lower fertility of the better-educated can only mean that dysgenic processes have been present.

A further study using a different methodology to demonstrate a significant heritability of educational attainment has been published by Teasdale and Sorensen (1983). They obtained information on years of education of 1,417 adopted children in Denmark and examined this in relation to the socioeconomic status of their biological fathers. For males the correlation was .20 and for females .22. The correlations with the socioeconomic status of the adopted fathers were .33 (males) and .32 (females). The positive and significant correlations between the educational attainment of adopted children with the socioeconomic status of their biological fathers suggests genetic transmission of the qualities determining educational attainment.

9. CONCLUSIONS

There are three principal points of interest in the studies of the relationship between educational level and fertility reviewed in this chapter. First, the relationship has been extensively investigated in many countries and has been invariably found to be inverse. Because educational attainment is positively associated with intelligence at a correlation of around .6 (Eysenck, 1979; Brody, 1992), the inverse relationship between educational attainment and fertility corroborates the evidence on the inverse relationship between intelligence and fertility reviewed in Chapters 5, 6 and 7. We noted in Chapter 6 that there were some American studies carried out in the 1960s that suggested fertility had become eugenic, but subsequent and better-sampled studies of the 1980s showed this was not so, and we concluded that the later studies were more sound. The inverse and substantial relationship between educational attainment and fertility confirms this conclusion. Indeed, considering the strong relationship between educational attainment and intelligence, it is impossible that fertility for intelligence could be eugenic at the same time that fertility for educational attainment has been dysgenic. Our first inference from the extensive evidence on the inverse relationship between educational attainment and fertility is that it confirms the conclusion that fertility for intelligence has been dysgenic in Western nations from the last decade of the nineteenth century and the whole of the twentieth century.

The second feature of interest in the evidence reviewed in this chapter is that dysgenic fertility in relation to educational attainment was generally stronger among the earlier birth cohorts born in the closing decades of the nineteenth century and first decades of the twentieth and progressively slackened as the twentieth century unfolded. Nevertheless, there are no countries in which dysgenic fertility has entirely disappeared, and anticipated fertility among relatively youthful cohorts born in the second half of the twentieth century in the United States, Britain and The Netherlands remains dysgenic. This general trend for dysgenic fertility to slacken over time is consistent with the evidence on the inverse relationship between intelligence and fertility reviewed in Chapters 5 and 6, especially Van Court and Bean (1985).

Third, although educational attainment is a fairly strong correlate of intelligence, it is also determined by motivational and personality traits responsible for sustained work, self-discipline, the capacity to work for long-term goals and so on. The trend for well-educated women to have comparatively few children suggests that fertility has been dysgenic for these motivational and personality traits as well as for intelligence. We shall examine this possibility more fully in Chapters 12 through 14. But before doing so we have to consider fertility differentials in relation to socioeconomic status as a further expression of dysgenic fertility.

Chapter 10

Socioeconomic Status and Fertility

1. United States. 2. England. 3. Australia and Canada. 4. Europe. 5. Japan. 6. Conclusions.

Socioeconomic status differences in fertility provide a further source of information on whether, and for how long, fertility in Western nations has been dysgenic. Like educational attainment, socioeconomic status is partly determined by intelligence, with which it is correlated at around .5 (Jencks, 1972; Brody, 1992). Hence, the relationship between socioeconomic status and fertility should be consistent with that between educational attainment and fertility and intelligence and fertility, that is to say, inverse. In this chapter we shall see if it is.

One of the points of examining the relationship between socioeconomic status and fertility is that it provides a further check on the previous results indicating that fertility has been dysgenic in Western nations for several generations. There are some countries for which data on fertility in relation to socioeconomic status go back to the early decades of the nineteenth century, making it possible to assess when dysgenic fertility began and how long it has been in place.

There is also a problem with the data on socioeconomic-status differences in fertility because they are mostly derived from censuses and generally consist of the number of children of married women in relation to the socioeconomic status of their husbands. They rarely provide data on the fertility of unmarried men or women. Dysgenic fertility is substantial among the unmarried because low socioeconomic-status women have higher unmarried fertility than high socioeconomic-status women. This was not too much of a problem in the nineteenth century and the first two-thirds of the twentieth because illegitimacy was below 5 percent among whites in the United States, in Britain and throughout most Western countries, so the omission of illegitimate children did not greatly affect the figures. However, from around 1970 the proportion of children born out of wedlock began to increase. In the United States, the illegitimacy rate among white women reached 10 percent by 1980 and 22 percent by 1991. Among American blacks, illegitimacy has always been rather higher. In 1950 it was 19 percent and had risen to 39 percent by 1980 (Murray, 1984, p. 126; Rubinstein, 1994). With figures like these the number of children of married women in relation to their husbands' socioeconomic status is not a satisfactory measure of the magnitude of dysgenic fertility. In Britain illegitimate births remained steady at around 4 to 5 percent of all births from 1870 to around 1960. Thereafter they began to increase, reaching 8 percent in 1970, 10 percent in 1980 and 26 percent in 1989 (Coleman and Salt, 1992, p. 135). Similar increases in illegitimacy have occurred from the 1960s in much of Continental Europe including France, Austria, Norway, Sweden and Denmark (Cooper, 1991). The effect of these increases has been that from around 1970, the number of children of married women in relation to the socioeconomic status of their husbands has ceased to provide reliable information on the degree to which fertility is dysgenic. Nevertheless, this information did provide useful information on the extent of dysgenic fertility when illegitimacy was low from the second half of the nineteenth century to around 1970, and a consideration of this evidence is our concern in this chapter.

1. UNITED STATES

Information on socioeconomic status and fertility in the United States was obtained in the censuses of 1900, 1910 and 1960 and in a survey carried out by the Bureau of the Census in 1990, the results of which have been reported by Bachu (1990, 1991). The results are shown in Table 10.1. The first four rows give the fertility of white wives, born in the years 1830–1840, 1861–1865, 1916–1925 and 1946–1954, in relation to their husbands' socioeconomic status. Notice that there is a relatively small dysgenic fertility ratio of 1.23 in the first cohort, an increase in the second cohort to 1.63, followed by falls to 1.25 and 1.10 in the third and fourth cohorts. This shows the increase of dysgenic fertility in the early years of the demographic transition followed by a decrease in dysgenic fertility in the later stage. Row 5 shows the fertility of white women born 1946–1954 in relation to their own occupation rather than that of their husbands and for whom dysgenic fertility is greater than for married women. This reflects the inclusion of unmarried women, who are more numerous in the white-collar occupations. The sixth row gives the fertility of men born 1916–1925 in relation to their socioeconomic status and also shows a dysgenic trend.

The fertility of blacks in relation to socioeconomic status is shown in rows 7 through 10. Row 7 shows pronounced dysgenic fertility of 1.59 among the 1916–1925 cohort, falling to 1.37 among the 1946–1954 cohort. Notice that for both cohorts dysgenic fertility is greater among black women than among whites. Row 9 gives the fertility of black women born 1946–1954 and again shows a greater dysgenic fertility than the same birth cohort of white women

Table 10.1
Fertility in Relation to Socioeconomic Status in the United States

Sample	Date of Birth	Age	Socio-economic Status 1	2	3	4	5	6	7	Dysgenic Ratio	Data	Reference
White wives	1830-40	60-70	5.6	5.1	7.8	5.6	7.1	6.9	5.2	1.23	1900 Census	Haines, 1992
White wives	1861-65	45-49	3.0	3.5	3.2	4.3	4.5	4.9	5.2	1.63	1910 Census	Kiser, 1970
White wives	1916-25	35-44	2.4	2.4	2.3	2.6	2.7	3.0	4.0	1.25	1960 Census	Preston, 1974
White wives	1946-54	35-44	2.0	2.0	2.0	2.2	- -	2.2	2.5	1.10	1990 Survey	Bachu, 1990
White women	1946-54	35-44	1.6	1.8	2.1	1.8	- -	2.2	- -	1.37	1990 Survey	Bachu, 1991

White men	1916-25	35-44	2.2	2.3	2.1	2.6	2.7	3.0	4.0	1.36	1960 Census	Preston, 1974
Black wives	1916-25	35-44	2.2	2.6	2.5	3.0	3.2	3.5	5.1	1.59	1960 Census	Preston, 1974
Black wives	1946-54	35-44	1.9	2.4	2.7	2.5	- -	2.6	3.9	1.37	1990 Survey	Bachu, 1990
Black women	1946-54	35-44	1.9	2.2	2.7	2.6	- -	2.9	- -	1.53	1990 Survey	Bachu, 1991
Black men	1916-25	35-44	1.9	2.3	2.1	2.8	2.9	3.1	4.3	1.63	1960 Census	Preston, 1974
Hispanic wives	1946-54	35-44	2.4	2.6	2.4	2.8	- -	2.9	3.6	1.21	1990 Survey	Bachu, 1990
Hispanic women	1946-54	35-44	2.3	2.3	2.8	2.6	- -	2.7	- -	1.17	1990 Survey	Bachu, 1991

Note: Socioeconomic class: 1 = professional; 2 = managerial; 3 = clerical; 4 = skilled; 5 = semiskilled; 6 = unskilled; 7 = farm workers.

(1.53 compared with 1.37). Row 10 shows dysgenic fertility among black men born 1916–1925 and that this is greater than the same cohort of whites. All the data show greater dysgenic fertility among blacks than among whites, consistent with the results for intelligence reviewed in Chapter 6.

The last two rows give fertility for Hispanic wives in relation to their husbands' socioeconomic status, and for Hispanic women in relation to their own occupational status. In both cases the trends are moderately dysgenic.

2. ENGLAND

The English census of 1911 contained information on the number of children of married women and the socioeconomic status of their husbands, and the number of children and socioeconomic status of the respondents' parents. The information contained in the census has been analyzed by Haines (1989) to calculate fertility differentials in relation to socioeconomic status for women born as early as 1815 to 1825 and for a later cohort born between 1865 and 1870. Fertility differentials in relation to socioeconomic status have been calculated by Benjamin (1966) for women born between 1900 and 1909 from the Royal Commission on Population of 1944, and for three cohorts of women born between 1921 and 1940 from the 1971 census by Coleman and Salt (1992). The whole set of data is shown in Table 10.2.

There are two principal features of interest in these figures. First, the dysgenic fertility ratios show that fertility has been dysgenic throughout the entire period from 1815 to 1940, a period of 125 years spanning five generations. Second, the extent of the dysgenic fertility is quite small in the first cohort born in the years 1815–1825; increases to its maximum in the cohort born between 1900 and 1909, when the unskilled working class had almost twice the number of children as the professional class; and then declined again until it had almost, but not entirely, disappeared in the last cohort between 1936 and 1940. This repeats the American data shown in Table 10.1.

Third, we note in the American data that the inverse socioeconomic-fertility differentials were initially linear but subsequently turned J shaped among later cohorts. The same phenomenon appears rather more clearly in the British data from 1921 onward, where fertility declined from class 1 to class 3 nonmanual and then increased to class 5. For some reason the emergence of a J-shaped relationship is more pronounced in Britain than in the United States.

3. AUSTRALIA AND CANADA

Data for Australia on occupation of husbands and fertility of married women were collected in the censuses of 1911 and 1921, and the results have been analyzed by Wong (1980). Number of children were not categorized by socioeconomic status but by occupation, such as professionals and public servants, construction workers, farm workers and so forth. These can be allocated ap-

Table 10.2
Completed Fertility of Married Women in Relation to Socioeconomic Status in England

Date of Birth	Socio-economic Status 1	2	3NM	3M	4	5	Dysgenic Ratio	Source	Reference
1815-25	6.90	7.50	- -	7.10	8.00	8.60	1.25	1911 Census	Haines, 1989
1865-70	2.60	3.40	- -	4.20	4.50	4.80	1.85	1911 Census	Haines, 1989
1900-09	2.33	2.64	2.37	2.89	3.96	4.45	1.91	1946 Survey	Benjamin, 1966
1921-25	2.04	1.99	1.86	2.20	2.24	2.47	1.21	1971 Census	Coleman and Salt, 1992
1926-35	2.18	2.09	1.95	2.29	2.32	2.61	1.20	1971 Census	Coleman and Salt, 1992
1936-40	2.23	2.12	2.00	2.29	2.31	2.58	1.16	1971 Census	Coleman and Salt, 1992

Note: Socioeconomic class: 1 = professional; 2 = managerial; 3NM = minor nonmanual; 3M = skilled; 4 = semiskilled; 5 = unskilled.

proximately to the conventional socioeconomic-status categories (1 = professional, senior executive, and so on), although all blue-collar workers have to be grouped into a single category; and farmers and farm workers are presented as a single group. The results are shown in Table 10.3. Notice that there was virtually no dysgenic fertility in the earliest birth cohort born before 1846 but appreciable dysgenic fertility among the cohort born 1877–1881, the latest cohort for which completed fertility could be calculated from the 1921 census.

In Canada the first data on social-class differences in fertility were collected in the 1941 census. They have been analyzed by Charles (1949) as age-standardized average number of children of married women and are shown for the five standard socioeconomic classes in Table 10.3. Fertility is strongly dysgenic.

Table 10.3
Socioeconomic-Status Differentials in Fertility in Australia and Canada

Country	Date of Birth	Socio-economic status 1	2	3	4	5	Farmers, Farm Workers	Dysgenic Ratio	Reference
Australia	1840-1846	6.26	6.58	- -	6.52	- -	7.76	1.04	Wong, 1980
	1877-1881	2.73	2.93	- -	3.21	- -	3.83	1.17	Wong, 1980
Canada	1860-1924	2.16	2.48	2.86	3.47	4.16	4.29	1.93	Charles, 1949

4. EUROPE

Fertility differentials in relation to socioeconomic status are shown for seven European countries for various cohorts in Table 10.4. The socioeconomic-class categories are broadly the same as those used for England. They exclude farm laborers, whose fertility is virtually always the highest of any social group. In some cases data are only available for the manual class as a whole, that is, for Czechoslovakia and Portugal, while in Norway no differentiation is made between skilled and semiskilled manual workers. The dates of birth are in some cases estimates from the dates of marriage and are therefore approximates.

The general features of the data resemble those we have already seen for the United States, England, Australia and Canada. There are four principal points of interest. First, in all cases the overall relationship between fertility and socioeconomic status is dysgenic with higher fertility among the unskilled working class than in the professional and middle classes. Second, in France where there are data for three cohorts, we see a slight rise in dysgenic fertility followed by a fall, reflecting the results found in the United States and England. Third, in Norway for which there are four data sets spanning the birth cohorts from 1885 to 1890 through 1925 to 1930, we see a steady decline in dysgenic fertility and the emergence of the J relationship from 1895 to 1910 onward that we observed in the United States and England. Fourth, in the data for Ireland and Northern Ireland there is a breakdown of fertility differentials for Catholics and Protestants showing higher overall fertility among Catholics. The reason for this is probably that Protestants use contraception more, and this is especially true for the professional and middle classes. Even among Catholics, however, there was significant dysgenic fertility among the birth cohorts of 1921 to 1925.

Table 10.4
Socioeconomic-Status Differentials in Fertility in Various European Countries

Country	Date of Birth	Socio-economic status 1	2	3	4	5	Dysgenic Ratio	Reference
Czechoslovakia	1931-35	1.42	1.83	--	2.20	--	1.55	Glass, 1967
France	1851-60	3.10	3.60	3.80	4.20	4.60	1.48	Haines, 1992
France	1861-70	2.60	2.90	3.20	3.70	3.90	1.50	Haines, 1992
France	1940-49	2.20	2.30	2.30	2.50	3.10	1.41	Desplanques, 1988
Ireland (Catholics)	1921-25	3.63	3.88	4.45	4.70	4.71	1.30	Glass, 1967
Ireland (Protestants)	1921-25	2.15	2.90	2.59	2.86	3.39	1.58	Glass, 1967
Northern Ireland (Catholics)	1921-25	3.70	4.01	4.53	4.09	4.47	1.21	Glass, 1967
Northern Ireland (Protestants)	1921-25	1.96	2.26	2.27	2.41	2.70	1.37	Glass, 1967
Netherlands	1920-25	3.04	3.41	3.28	3.64	4.02	1.32	Glass, 1967
Norway	1885-90	3.80	5.28	5.91	5.99	6.07	1.60	Haines, 1992
Norway	1895-00	3.52	3.45	3.93	--	4.72	1.34	Glass, 1967
Norway	1915-20	2.65	2.53	2.36	--	3.33	1.25	Glass, 1967
Norway	1925-30	2.57	2.62	2.39	--	2.62	1.02	Glass, 1967
Portugal	1930-40	2.69	3.02	--	3.37	--	1.25	Ashurst et al., 1984

5. JAPAN

The Japanese censuses of 1950 and 1960 have been analyzed for socioeconomic-status differences in the fertility of married women aged 45 and over by Taeuber (1960). Three occupational groupings were used consisting of white-collar workers (class 1), manual workers (class 2) and farmers and fishermen (class 3). The results are shown in Table 10.5. Notice that fertility is dysgenic for both birth cohorts.

Table 10.5
Number of Children of Married Women in Japan Analyzed by Occupations of Husbands

Date of Birth	Non-manual	Manual	Farmers, Fishermen
1870-1905	3.56	3.79	5.06
1880-1915	3.37	3.82	4.02

Source: Taeuber (1960).

6. CONCLUSIONS

In this chapter we have seen that there was an inverse relationship between socioeconomic status and fertility in many Western nations among cohorts born in the early and middle decades of the nineteenth century, and that this continued into the twentieth. The magnitude of the inverse relationship was relatively low in the first half of the nineteenth century, became larger in the second half and early decades of the twentieth century and then began to narrow again. These trends are apparent in the time series data for the United States, England, France and Norway. The slackening of dysgenic fertility in relation to socioeconomic status in the later decades of the twentieth century is consistent with the evidence on dysgenic fertility in relation to intelligence reviewed in Chapters 5, 6 and 7, and in relation to educational attainment reviewed in Chapter 9. All the evidence hangs together. However, there are no countries in which dysgenic fertility has entirely disappeared.

Although the overall relationship between socioeconomic status and fertility is consistently inverse, there are some instances where it has become J shaped, with moderate fertility among the professional class declining to lower fertility among clerical workers, and then increasing to high fertility among semiskilled and unskilled workers. This relationship was present in the United States and England, and there is a hint of it in Norway. This probably reflects efficient family planning by the middle classes, among whom the higher professional classes have more children because they can afford more. The rising fertility among the three working-class groups probably reflects increasingly inefficient use of contraception and unplanned births with declining socioeconomic status.

In addition to confirming the presence of dysgenic fertility in modern populations, the inverse relationship between socioeconomic status and fertility shows that this was present in the early and middle decades of the nineteenth century in England, the United States, Australia, Canada and France. It appears, therefore, that dysgenic fertility has been present for around six generations, from the 1815–1825 birth cohorts, assuming the data for England are generalizable to other Western nations, to the birth cohorts of the 1950s–1960s.

It is difficult to estimate with any precision the extent of the deterioration of genotypic intelligence arising from dysgenic fertility over six generations. There is a problem because the magnitude of dysgenic fertility varies at different points in time during the course of the nineteenth and twentieth centuries, being initially low, increasing and then declining again. In Chapter 5 it was proposed that the evidence on the association between intelligence and number of siblings in England indicated a deterioration of five IQ points between 1890 and 1980. If this is projected backward in time to the 1815–1825 cohorts, there would have been a further decline of about three IQ points, making an eight IQ point decline since the first appearance of dysgenic fertility in the early nineteenth century. The American data suggests that the decline has been a little less. In Chapter 6 it was proposed that the decline of genotypic intelligence in the United States for the three generations from the birth cohort of 1890–1894 to the second half of the twentieth century was about 2.50 IQ points. The socioeconomic-status and fertility evidence reviewed in this chapter, showing the presence of an inverse relationship from the birth cohorts of 1815 onward, indicates a further three generations of dysgenic fertility. It seems reasonable to assume that this was of about the same magnitude, giving a deterioration of genotypic intelligence over the six generations of around five IQ points. Thus, we have two estimates of the magnitude of the deterioration of genotypic intelligence in Western nations since the early decades of the nineteenth century of eight and five IQ points.

Chapter 11

Socioeconomic-Status Differences in Intelligence

1. Socioeconomic Status and Intelligence. 2. The Intergenerational Transmission of Intelligence. 3. Intelligence and Social Mobility. 4. Genetic Differences between the Social Classes. 5. Heritability of Social-Class Differences in Intelligence. 6. Conclusions.

In the last chapter, we saw that there has been an inverse relationship between socioeconomic status and fertility in Western nations since the early and middle decades of the nineteenth century. It was assumed by Francis Galton and his fellow eugenicists that there was a genetically based social-class gradient for intelligence, such that the professional class had the highest intelligence level and that intelligence declined progressively among the middle, lower middle, skilled artisan and unskilled social classes. This is why they were worried about the inverse association between social class and fertility. The belief that the social classes differ genetically in respect of intelligence has not been allowed to pass unchallenged. In the second half of the twentieth century, many sociologists asserted, or even took it for granted, that the social-class differences in intelligence were entirely determined by environmental advantages and disadvantages. If this is so, there is no need for concern about the low fertility of the professional and middle classes in comparison with the working classes. The eugenicists' concern on this issue depends crucially on the demonstration that the social classes do differ genetically in regard to intelligence.

The eugenic case that socioeconomic-status differences in fertility are dysgenic consists of the propositions that social-class differences in intelligence exist; that they are genetically based; that this explains the transmission of intelligence in social classes from parents to children; and that the genetic basis of social-class differences in intelligence is reinforced in each generation by the process of social mobility. We examine these propositions in this chapter.

Table 11.1
Mean IQs of American Adults in Relation to Their Social Class

Date	Number	Social Class						Reference
		1	2	3	4	5	Farm workers	
1918	28,597	123	119	104	98	96	97	Johnson, 1948
1941-45	18,782	120	113	106	96	95	94	Johnson, 1948
1981	1,880	111	104	99	93	89	--	Reynolds et al., 1987

Note: Class: 1 = professional; 2 = clerical; 3 = skilled; 4 = semiskilled; 5 = unskilled.

1. SOCIOECONOMIC STATUS AND INTELLIGENCE

The first eugenicists like Galton believed that the professional class was the most intelligent and that intelligence declined progressively in the middle, skilled, semiskilled and unskilled lower class. They had no intelligence tests with which they could demonstrate this belief. But with the development of the intelligence test by Alfred Binet in France in 1905, it became possible to collect data on this issue. This was to show that there is indeed a socioeconomic gradient for intelligence of the kind the eugenicists supposed.

Studies of this kind divide the population into social classes and calculate the mean IQ for each class. The results of three American studies of this kind are shown in Table 11.1. They are derived from military personnel in World War I and World War II and from the 1981 standardization sample of the Wechsler Adult Intelligence Scale Revised, and they span a period of 63 years. All three studies show that the professional class has the highest average IQ, and that intelligence falls steadily with decline in socioeconomic status. It may be noted that in the first two studies the average IQs are higher than in the later study. The reason for this is probably that the professional class was a small elite at the time of the earlier studies and had grown considerably in size; hence, it became diluted by the time of the third study. Conversely, the semiskilled and unskilled were more numerous and contained a greater reservoir of intelligence in the two earlier studies, whereas by 1981 these classes had shrunk and their intelligence level had fallen.

Another way of expressing the relationship between social class and intelligence is by means of the correlation between them. The literature on this issue has been reviewed by Brody (1992) who arrives at a correlation of .50. Correlations of this size are sometimes criticized as being quite low, but should be adjusted for the imperfect reliability of the measures of intelligence and social class. Any particular measure of intelligence and social class is not perfectly reliable because measures taken on subsequent occasions sometimes give different results. For instance, in the British social-class categories, an optician is counted as class 1 and a member of parliament as class 2, so an individual's social class can change with a change of career. IQs are also not perfectly reliable, due to measurement error. IQs can be credited with a reliability of .9, following Bouchard (1993). The reliability of social class can be calculated from a study by Teasdale and Owen (1984a) of 372 adopted men in Denmark whose socioeconomic status was assessed between the ages of 25 and 28 and again between the ages of 35 and 38. The correlation between the two was .75, and this provides a measure of the reliability of socioeconomic-status measures. Correcting the correlation of .50 for these two reliability coefficients gives a correlation of .60. This is a better estimate of the true correlation between socioeconomic status and intelligence.

The reason that this correlation is not higher is that social class is partly determined by personality traits, such as motivation and the capacity for hard work, and sometimes by mental disorders. Thus, an individual of relatively modest intelligence of around 100 can get into social class 1 by working hard, and conversely another individual with the same IQ who is unable to work hard, or perhaps suffers from a mental disorder, can end up in social class 5. Nevertheless, the association between social class and intelligence is quite substantial, as the eugenicists supposed.

2. THE INTERGENERATIONAL TRANSMISSION OF INTELLIGENCE

There is not only a positive association between people's socioeconomic status and their intelligence, but also a positive association between people's socioeconomic status and the intelligence of their children. A summary of 102 studies on this issue has been made by White (1982), who concludes on the basis of a meta-analysis of these that the correlation is .407. This figure needs correction for unreliability; using the coefficients of .90 and .75, derived in the last section, raises the correlation to .49. The correlation between socioeconomic status and intelligence among children is thus a little lower than the correlation among adults, which we estimated at .60. The reason for this lies in regression effects among children, that is, the children of high-intelligence parents, typically in social class 1, are, on average, a little less intelligent than their parents. Conversely, the children of low-intelligence parents, typically in social class 5, are, on average, a little more intelligent than their parents.

Some illustrative data on the relation between parents' socioeconomic status and their children's average IQs, drawn from Britain, France, Russia and the United States, are shown in Table 11.2. The British, French and Russian data

Table 11.2
Mean IQs of Children by Socioeconomic Status of Fathers

Country	Date	Socioeconomic status						Reference
		1	2	3	4	5	Farm workers	
Britain	1920s	115	113	104	98	95	96	Duff and Thomson, 1923
Britain	1960s	112	106	101	- -	97	- -	Douglas, 1964
France	1950s	120	106	- - -	99	- -	93	Zazzo, 1960
France	1974	138	121	109	90	85	85	Olivier and Devigne, 1983
Russia	1920s	117	107	101	97	92	- -	Sirkin, 1929
United States	1930s	113	106	102	94	94	- -	Terman and Merrill, 1937
United States (whites)	1970	108	102	99	96	91	- -	Kaufman and Doppelt, 1976
United States (blacks)	1970	88	87	84	83	78	- -	Kaufman and Doppelt, 1976

came from large surveys of several thousand children and the American data from the standardization samples of the Stanford-Binet and the Wechsler Intelligence Scale for Children. The results indicate a difference of around twenty IQ points between the children of the professional class and those of unskilled workers.

The bottom row of Table 11.2 shows that IQs of black children are related to the socioeconomic status of their parents in the same way as white children, although the IQs are all depressed downward by around fourteen IQ points. This shows that, even within socioeconomic-status categories, the IQs of black children are markedly lower than those of white children.

Because parents' socioeconomic status is associated with their intelligence, and also with the intelligence of their children, the intelligence of parents and children is also correlated. The correlation is often said to be about .5, but this is for one parent and one child. The more important correlation is for both parents and all their children, and this is significantly higher. Bouchard (1993,

p. 54) gives the median value from all existing studies as .73. This also needs to be corrected for unreliability to give a true correlation of .81.

3. INTELLIGENCE AND SOCIAL MOBILITY

The eugenicists did not believe that the socioeconomic classes constitute a rigid caste system in which individuals remain for life in the social class into which they are born. On the contrary, they recognized that many people move out of their parents' class and into another, and they do this in accordance with their talents, among which intelligence is an important component. Subsequent research has shown that this was right. On average, children have similar intelligence to that of their parents, subject to a small amount of regression to the mean, and hence appropriate for the social class into which they are born. However, this is only an average. Individual children frequently have IQs significantly higher or lower than the average of their parents, and consequently incongruent with their parents' socioeconomic status. When this happens, these children typically move upward or downward, into the social class for which their intelligence fits them.

This is the process of social mobility through which children in each generation are reallocated to the social class for which their intelligence is appropriate. For instance, children with average IQs born to low social-class families tend to be upwardly mobile into the middle range of the socioeconomic-status hierarchy, while such children born into professional class families tend to move downward. A useful way of demonstrating this process is by taking pairs of brothers because this provides controls for family and school influences. Brothers are often born with different IQs, and those with the higher IQs tend to move socially upward and those with lower IQs move downward. This effect has been demonstrated and quantified by Jencks (1972) who calculated that a fifteen IQ point difference between the IQs of brothers in adolescence produced a 17 percent difference in their earnings. The brother with the higher IQ secures higher earnings.

Social mobility works largely through the educational system. Whatever their social origins, intelligent children tend to do well in school, proceed to college and enter the professional and managerial classes. Intelligence is an important determinant of educational attainment, with which it is correlated at a magnitude of around .6. Thus intelligence, through its effect on educational attainment, becomes a determinant of socioeconomic status.

The extent to which social mobility across social classes occurs is measured in studies of the percentage of children who end up in a different social class from that of their parents. A number of studies of this kind of the extent of social mobility in nineteenth-century America and Europe have been collated by Kaelble (1985). He uses a four–social class system consisting of the middle, lower middle, skilled and unskilled working class. He estimates that in terms of this system, 52 percent of sons were socially mobile in three American cities,

Table 11.3
Percentage Manual/Nonmanual Mobility

Canada	37.5
Sweden	37.0
USA	36.5
France	34.0
England & Wales	33.7
Australia	33.2
Poland	30.7
Bulgaria	28.5
Japan	28.0
Italy	25.5

Source: Heath (1981).

and 46 percent were socially mobile in twelve European cities in Austria, Britain, Denmark, France, Germany and The Netherlands. Evidently there was an appreciable amount of social mobility in the nineteenth century in which the talented moved up the social hierarchy and the less talented moved down. For the twentieth century, studies of mobility across the working class–middle class divide for a number of nations have been collated by Heath (1981) for the early post–World War II decades. His figures are shown in Table 11.3. It will be seen that about a third of sons are socially mobile in this set of nations, ranging from a low of 25.5 percent in Italy to a high of 37.5 percent in Canada.

If more social classes are used in these analyses, the extent of social mobility becomes greater because with more classes people have a greater probability of ending up in a different class from their parents. This can be illustrated by a study in Britain by Goldthorpe, Llewellyn and Payne (1987) on approximately 10,000 men. They used both a three-division and a seven-division social-class classification. The three-division classification consisted of the upper class (professional and managerial), middle (lower white-collar and skilled working) and lower classes (semiskilled and unskilled). The results were that 59 percent of the sons of upper-class men remained in the upper class, 26 percent fell into the middle class and 15 percent fell into the lower class. Conversely, among those born into the lower class, 57 percent rose into the middle class and 16 percent rose into the upper class.

If the social hierarchy is divided into seven classes, more social mobility appears to result. For instance, 46 percent of boys born into professional-class families end up in the professional class themselves while 7 percent of those born in the lowest class join the professional class. Thus, boys born into the professional class are about six to seven times more likely to end up in the professional class themselves as compared with boys born into the lowest social class. These studies have confirmed the eugenicists' belief that the social classes tend to perpetuate themselves from generation to generation, but that this is mitigated by some social mobility across social classes.

4. GENETIC DIFFERENCES BETWEEN THE SOCIAL CLASSES

Both genetic and environmental factors are responsible for the tendency of about two-thirds of children to remain in the social class into which they are born, and for the remaining third to be socially mobile out of their class of origin. The genetic contribution is that the social classes have come to be, to some degree, differentiated for the genes and alleles (alternative forms of a gene) responsible for intelligence. The genes for high intelligence have become disproportionately represented in the higher social classes, particularly in the professional class, while those for low intelligence have become disproportionately represented in the lower classes. However, such is the complexity of the genetical processes determining intelligence that the segregation of the genes is not complete. The professional classes continue to carry some genes for low or moderate intelligence; conversely, the lower classes have some genes for moderate to high intelligence. The partial but incomplete segregation of the genes for intelligence into the social classes is one of the principal reasons for the intergenerational continuity of social class, tempered by social mobility in those cases where children inherit an intelligence level different from that of their parents. There are also two other factors operating. One of these consists of the character, personality and motivational traits which also determine social class. They operate in the same way as intelligence and will be considered in Chapters 12 through 14. The second consists of environmental factors through which parents generally attempt to socialize their children into their own value system, professional and middle-class parents endeavouring to foster their distinctive class values in their children. Research usefully summarized by Rowe (1994), however, indicates that these efforts are not very successful and that intergenerational social-class continuities are principally genetic.

5. HERITABILITY OF SOCIAL-CLASS DIFFERENCES IN INTELLIGENCE

Herrnstein's syllogism stated that where intelligence has a high heritability and intelligence is a major factor in social mobility, there must be genetic differences between the social classes. Research on adopted children provides direct evidence that this is the case. The general format of this research has been to show that children separated from their fathers shortly after birth nevertheless differ in intelligence according to the social class of the fathers. A professional-class father has a more intelligent child, whom he has never seen, than an unskilled working-class father.

The first study to provide evidence of this kind was carried out in England in the 1920s by Jones and Carr-Saunders (1927). They investigated the IQs of orphanage children in relation to the occupational status of their biological fathers. The results were that the children of professional-class fathers had an average IQ of 107, and the children's IQs fell with fathers' declining social class—to 93 for the children of laborers. Four years later a similar study was published in England by Lawrence (1931) which showed a correlation of .25 between the IQs of children reared partly in orphanages and partly by foster parents, and the social status of their biological fathers. These early workers perceived that the positive correlations they obtained between the socioeconomic status of biological fathers and their children reared in orphanages and by foster parents could only be explained by fathers of higher social status transmitting higher intelligence to the children; hence, socioeconomic-status differences must have some genetic basis.

Similar studies showing a positive relationship between the IQs of adopted children and the socioeconomic status of biological fathers were to appear a few years later in the United States. The first study was done by Leahy (1935) on 194 adopted children. The average IQ was highest at 113 of those whose fathers were professionals, and fell progressively to 112 (class 2), 111 (skilled and clerical), 109 (semiskilled) and 108 (unskilled). The next study to show this was carried out by Skodak and Skeels (1949) who obtained 70 adopted children for whom the social class was known of both the adoptive and the biological fathers. The correlations between the children's IQs and their adoptive and biological fathers have been calculated by Herrnstein (1971, p. 87) as −0.06 and .29, respectively. Note that in this study there is no correlation between the IQ of a child and the social class of the adopted father, showing that social class has no environmental effect on the IQ. There is, however, a significant correlation between the IQ of an adopted child and the social status of the biological father, showing that the higher intelligence of children of higher social-status fathers must have been transmitted genetically.

The next American study to confirm this conclusion was carried out by Munsinger (1975) on 41 adopted children in California. The correlations of the children's IQ were −0.14 with the adopted parents' social status, and .70 with the biological parents' social status. This latter correlation is high, but given the small sample size barely differs significantly from the correlations of .25 and .29 found in the earlier studies. Notice again the absence of any association between the children's IQs and the social status of their adoptive parents.

These results were confirmed again in a study by Weinberg, Scarr and Waldman (1992) on black, interracial (children of one black and one white parent) and white children adopted by white couples. The authors present some correlations between the IQs of the black and interracial children (those for whites are curiously omitted) at the age of 17 and the education and occupation of their biological and adoptive fathers. The correlation with the adoptive fathers' occupational status was .03, confirming the zero correlations found in the earlier studies. But the correlations with the biological fathers' and mothers' education, a good proxy for occupational status which was not available, were .28 and .23, about the same as those found in the earlier studies. Once again the only way to explain the results is that higher social-class parents must have genetically based higher IQs, which they transmit to their children.

The final study of this kind comes from France. Capron and Duyme (1989) report results for eighteen adopted children of middle-class fathers and twenty adopted children of working-class fathers. The children of the middle-class fathers had a mean IQ of 114 and those of working-class fathers of 98. The only explanation for this difference must be that the children of the middle-class fathers inherited higher IQs than those of the working-class fathers.

6. CONCLUSIONS

The eugenicists believed that the professional and middle classes were genetically superior to the lower classes in respect to intelligence. As Lewis Terman (1922) wrote in connection with the socioeconomic-status differences in intelligence that had begun to appear in the early work on intelligence testing: "The children of successful and cultivated parents test higher than children from wretched and ignorant homes for the simple reason that their heredity is better" (p. 671).

We have seen in this chapter that the research evidence supports this conclusion. Through the process of social mobility, individuals with genes for high intelligence tend to move up in the class hierarchy into the professional and middle classes, while those with the genes for low intelligence tend to move down into the unskilled working class and the underclass. This has led to the genes for intelligence becoming to some degree segregated by social class. As Ronald Fisher expressed it in the heyday of eugenics, "social class thus became genetically differentiated by the agencies controlling promotion or demotion" (1929, p. 245).

This genetic segregation of the social classes is not complete. There are still some genes for low intelligence in the professional classes, and some genes for high intelligence in the lower classes. Nevertheless, genetic segregation is sufficiently pronounced to produce a considerable degree of continuity in the transmission of high intelligence in the professional and middle classes from generation to generation, and in the transmission of low intelligence in the unskilled and underclass. This is the genetic basis for what has come to be called the "cycle of disadvantage," in which the unskilled working class and the underclass tend to reproduce themselves from one generation to the next, and the corresponding though less well-documented "cycle of advantage," in which high ability appears in certain families—such as the Darwins and the Rothschilds—over the course of generations. The eugenicists believed that the social classes have become to some degree genetically differentiated with regard to intelligence. In this chapter we have seen that they were right.

Chapter 12

Socioeconomic-Status Differences in Conscientiousness

1. Character and Conscientiousness. 2. Sociological Descriptions. 3. Moral Values. 4. The Work Ethic. 5. Smoking. 6. Alcoholism. 7. Sexual Restraint. 8. Crime. 9. Psychopathic Personality. 10. Conscientiousness and Social Mobility. 11. IQ × Conscientiousness = Achievement. 12. Conclusions.

The eugenicists were concerned that the breakdown of natural selection entailed a deterioration of the genetic quality of the populations of the Western nations for health, intelligence and the personality trait they described as *character.* We have completed our examination of the evidence for genetic deterioration since the early nineteenth century in regard to health and intelligence. We turn now to the third of the characteristics for which the eugenicists believed that genetic deterioration was taking place. This was for the personality qualities they called *character*, used by Galton to describe a syndrome of socially desirable qualities comprising honesty, integrity, prudence, self-discipline, the control of antisocial behavior, including crime; a sense of social obligation; and the motivation to work for long-term goals. Galton and his fellow eugenicists attached as much importance to character as they did to intelligence, and they thought that genetic deterioration in regard to character was equally serious.

In considering whether concern about genetic deterioration in regard to character was justified, there are three major issues that need to be addressed. The first concerns the nature of character and whether the eugenicists were correct in their belief that the professional and middle classes are superior to the lower classes in regard to character. The present chapter is concerned with this question. The second issue is whether the socioeconomic-status gradient for character has a genetic basis and will be considered in Chapter 13. Finally, in Chapter 14, we look at the problem of whether the populations of the Western nations are undergoing genetic deterioration in regard to character.

1. CHARACTER AND CONSCIENTIOUSNESS

We begin by clarifying the eugenicists' concept of character. The term has not been adopted in the vocabulary of psychology, but describes an important human attribute that has been recognized, studied and designated by a variety of different terms. In the writings of Sigmund Freud, the concept of the *superego* as the embodiment of the moral sense and conscience is closely similar to the idea of character and has been adopted by Cattell (1957) to describe one of the traits in his taxonomy of personality. Other terms employed by different theorists include *constraint* (Tellegen, 1985), *self-control* (Lorr, 1986) and *dependability* (Tupes and Christal, 1961). Eysenck's (1992) concept of *psychoticism* captures the poor socialization of those with weak character. The acceptance of work as a moral obligation and the capacity to work for long-term goals have been designated the work ethic (Weber, 1904) and achievement motivation (McClelland, 1961).

From the early 1980s there has been an emerging consensus among psychologists that *conscientiousness* is the best term for this trait. This term was first proposed by Norman (1963) and has become adopted by a number of leading personality theorists, such as Costa and McCrae (1988) and Brand (1995). It is the most satisfactory term for the concept because its meaning is well-understood in ordinary discourse, and it conveys the broad nature of the trait expressed over a wide range of values and behavior, including attitudes toward the law, work, personal health, personal relationships, social responsibility and moral principles. This term will therefore be adopted from now on to designate the trait the eugenicists described as character.

Conscientiousness has been most fully described by Paul Costa and his associates in textbooks by Costa and McCrae (1992) and Costa and Widiger (1994). It is envisaged as a broad personality trait of which the most important components are a well-developed moral sense embracing a strong social conscience, work ethic and self-discipline and self-control over antisocial, immoral and illegal temptations. Weak conscientiousness expresses itself in immoral, antisocial, personally irresponsible behavior—including alcoholism, drug abuse, and an antipathy to work. At the low extreme, conscientiousness is expressed in criminal behavior and in the psychiatric disorders of psychopathic personality and antisocial personality disorder. Our objective in this chapter is to show that in all these expressions of conscientiousness there is a socioeconomic-status gradient such that conscientiousness is strongest in the professional class and declines progressively through the other social classes.

2. SOCIOLOGICAL DESCRIPTIONS

From the beginning of the twentieth century sociologists studying social-class differences in values, attitudes and behavior observed that the professional and middle classes were stronger on what psychologists have come to call conscientiousness. An early British sociologist to note this was B.S. Rowntree (1901) who made an extensive study of poverty and its causes. He distinguished between what he called *primary poverty* and *secondary poverty*. Primary poverty was caused by low incomes in relation to expenditures and was mainly experienced by families in which there were large numbers of children who had to be maintained on low wages, and by the old who had no income or savings. Secondary poverty was caused by irresponsibility and fecklessness of various kinds, such as spending money on alcohol and tobacco that could not be afforded, or betting on race horses or dogs. Secondary poverty as Rowntree described it arose from an irresponsible use of money and was, in his judgment, a significant component of the poverty of the working classes.

Later in the century sociologists formulated the concept of *embourgeoisement* (Runciman, 1964). The idea was that the middle classes had strong bourgeois values of professionalism in their work, personal responsibility and thrift, values central to the concept of conscientiousness. The working classes, according to this view, were weaker in regard to these values. They tended to work simply for the money, were, to some degree welfare-dependent insofar as they lived largely in subsidized housing, and did not reckon to save for old age and other future contingencies. Sociologists debated whether, with the increasing affluence generated by economic growth, the better-paid working class would come to adopt middle-class values through the process of *embourgeoisement* and decided that they did not (Goldthorpe, Lockwood, Bechhofer and Platt, 1969).

Similar analyses were made in the descriptions appearing from the 1960s onward of what came to be known as "the underclass." This term was first coined by the American sociologist Oscar Lewis (1961) for a segment of society that does not subscribe to the conventional moral order of work and personal responsibility. The underclass, according to Lewis, has "a strong present-time orientation with little ability to delay gratification and plan for the future" (p. xxvi). The unwillingness or inability to delay gratification is central to this syndrome of values and was restated a few years later by the sociologist T.R. Sarbin (1970, p. 33) who observed that "the degraded poor are rooted in the present and are indifferent to the future." The underclass is a section of society in which conscientiousness is so low that it has become a social pathology.

3. MORAL VALUES

Strong moral values are an important component of conscientiousness and one in which socioeconomic-status differences have frequently been found. The first major study to demonstrate that the professional and middle classes have stronger moral values than the lower classes was carried out in the United States

Table 12.1
Median Rank Order of Four Values in Relation to Earnings

	Income						
	Low		Middle			High	
	1	2	3	4	5	6	7
A Sense of Accomplishment	10.4	10.3	9.1	9.4	8.4	7.6	6.1
Mature Love	14.4	14.0	12.3	12.2	10.8	11.5	11.8
A Comfortable Life	7.2	8.5	8.4	8.1	10.0	11.0	13.4
Pleasure	13.6	14.5	14.7	14.7	15.1	15.0	15.2

Source: Rokeach (1973).

in the mid-1960s by Kohn and Schooler (1969). They commissioned the National Opinion Research Center to interview a representative sample of 3,100 men and record replies to a number of statements. These included a set of five concerned with moral values, of which the first was "It's all right to do anything you want as long as you stay out of trouble." Scores on the mini-questionnaire showed a statistically significant correlation of .18 with socioeconomic status analyzed into the usual five-class categories.

In the 1970s this result was to be confirmed using an alternative methodology by Rokeach (1973). In this study a list of eighteen values was drawn up, and people were asked to rank them in order of the importance they attached to them. Four of them were moral values, namely *A Sense of Accomplishment* and *Mature Love* (positive moral values) and *A Comfortable Life* and *Pleasure* (negative moral values). People's rankings of these values were examined in relation to their income in a sample of 1,325 men and women, and it was found that high-earners attached greater value to *A Sense of Accomplishment* and *Mature Love*, while low-earners attached greater value to *A Comfortable Life* and *Pleasure*. The results are shown in Table 12.1, where a low figure indicates a high rank. Thus, the highest earners ranked *A Sense of Accomplishment* at 6.1, while the lowest earners ranked it much lower at 10.4. Hence, it appeared that low earners attached relatively little importance to *A Sense of Accomplishment* and

Mature Love. They attached greater value to *A Comfortable Life* and *Pleasure.* All the earnings differences in values are statistically significant.

A related approach has been the study of socioeconomic-status differences in what psychologists have called the capacity to "delay gratification," that is, the ability to restrain indulgence in immediate pleasure because of adverse consequences in the long term. For example, getting drunk or taking drugs today may be pleasurable, but we suffer for it the next day. Smoking cigarettes may be pleasurable now but we are likely to suffer for it when we develop cancer in middle age; socializing with our friends is pleasurable, but if we were studying or working instead, we would gain the benefit of a better education and career; spontaneous sex without the use of contraception may be pleasurable, but here again it is likely to have disadvantageous long-term consequences.

Self-control over the gratification of present pleasures in the interests of long-term advantage is the hallmark of the conscientious personality, and a series of studies by Mischel has shown that this capacity is stronger in the middle than in the working classes. Mischel (1958, 1961) had children choose between a small reward today or a larger reward at some time in the future. He found that middle-class children tended to prefer the larger reward in the future and working-class children the smaller reward today.

4. THE WORK ETHIC

A second important component of conscientiousness is a strong work ethic—a moral commitment to work and to the performance of work to a high standard. In the social sciences, the concept of the work ethic was first formulated by the German sociologist Max Weber in 1904. The essence of the idea was that people differ in the strength of their commitment to work. Those high in the trait work from a feeling of personal and social duty, while those low in the trait work solely to obtain money. Weber held that the work ethic was stronger in Protestantism than in Catholicism, and that this was an important reason that the Protestant nations of Northern Europe advanced more rapidly economically from the seventeenth century onward than the Catholic countries of Southern Europe, notably Spain, Portugal and Italy. Later it became recognized that a strong work ethic was not necessarily confined to the Protestant faith.

A number of studies have shown that the work ethic is stronger in the professional and middle classes than in the working classes. For instance, Abrams (1985) has reported the results of a social survey of attitudes among a representative sample of 1,231 adults in Britain. One of the questions was, Do you take a pride in your work? (put only to those in employment). The question was answered affirmatively by 84 percent of middle-class respondents, 78 percent of skilled workers and 70 percent of semiskilled and unskilled. Another question concerned whether importance was attached to one's job being socially valuable. This question was answered affirmatively by 49 percent of the middle-class respondents, 31 percent of skilled workers and 25 percent of unskilled.

A similar study by Mann (1986) concerns a study of responses by a representative sample of 1,769 adults in Britain analyzed by social class to the statement "Work is more than just earning a living." The percentage of respondents positively endorsing this question was highest among the professional and managerial class at 87 percent and fell progressively to 54 percent among the semiskilled and unskilled. Other questions in the same vein showed a similar socioeconomic-status gradient.

In the 1950s and 1960s the concept of work ethic was rediscovered by the American psychologists Atkinson (1958) and McClelland (1961). They renamed it "achievement motivation" and stripped it of its Protestant association, but apart from this modification the concept is both conceptually and empirically identical to Weber's original concept of the work ethic. The empirical similarity of the two concepts has been shown in several studies where questionnaire measures of the two traits have been shown to be highly correlated. I have found this in a 42-nation study in which various questionnaire measures of work ethic, achievement motivation and other work attitudes were administered and factor-analyzed; in all of these work ethic and achievement motivation were found to be correlated and expressions of a single factor (Lynn, 1991). As with the work ethic, studies of socioeconomic-status differences in achievement motivation have found generally higher scores in the professional and middle classes. Allen (1970) reviews five such studies, and further evidence has been summarized by Furnham (1990).

5. SMOKING

It has long been recognized that among our moral obligations is the duty to look after our health and not to abuse our bodies by self-indulgence of various kinds such as smoking, excessive alcohol consumption or drug abuse. Weakness in these regards is a moral failure and a further manifestation of low conscientiousness. Questionnaire measures of people's consciousness of the importance of health care have shown that it is a general trait associated with conscientiousness (Lynn, Devane and O'Neill, 1984), while at the extreme of low conscientiousness abuse of the body by drugs and alcohol is a well-known feature of the psychopathic personality. Because conscientiousness is stronger in the professional and middle classes than in the working classes, we should expect a socioeconomic-status gradient for these expressions of health care. Research has shown that this is the case.

As regards smoking, numerous studies in the United States and Europe have shown that it is inversely related to socioeconomic status (Flay, d'Avemas, Best, Kersell and Ryan, 1983; Conrad, Flay and Hill, 1992). Some illustrative figures derived from social surveys are shown for Britain, Denmark, Finland and Scotland in Table 12.2. Notice generally linear increases in the prevalence of smoking with declining social status, and the highest prevalence among the

Table 12.2
Percentages of Smokers by Socioeconomic Status in Various Countries

Country	Year	Sample Size	Sex	Socioeconomic status						Reference
				1	2	3	4	5	Unempl.	
Britain	1982	3,827	males	23	26	38	45	50	-	Blaxter, 1990
		4,974	females	21	26	35	37	45	-	Blaxter, 1990
Denmark	1985	3,468	females	27	26	34	44	48	52	Olsen and Frische,1993
Finland	1971	902	males	23	32	37	38	- -	-	Hassan, 1989
			females	20	27	- -	- -	- -	-	
Scotland	1992	827	males	- -	21	- -	41	- -	83	Glendinning, Shucksmith and Hendry, 1994
			females	- -	27	- -	36	- -	42	

Note: 1 = professional; 2 = clerical; 3 = skilled; 4 = semiskilled; 5 = unskilled.

Table 12.3
Percentages of Smokers by Education Level in the United States

	Years Education				
Sex	Below 12	13-15	16	17 plus	Reference
Females	27	--	--	12	Albrecht et al., 1994
Females	35	28	17	14	Williamson et al., 1989
Males	41	30	25	18	Winkleby et al., 1990
Females	36	24	15	17	Winkleby et al., 1990
Females	43	--	--	23	Matthews et al., 1989

unemployed in Denmark and Scotland, for which they were given as a separate category.

These inverse relationships between smoking and socioeconomic status are confirmed by several studies showing similar relationships between smoking and educational level. This is to be expected, since educational level is highly associated with socioeconomic status. Illustrative figures from four American studies are shown in Table 12.3.

The interpretation of these inverse relationships of smoking with socioeconomic status and educational level is that the higher levels of conscientiousness in the professional, middle and more educated classes act as a check on smoking. It is not, of course, the smoking as such that is a determinant of social mobility. Smoking is simply an expression of the strength of conscientiousness, which is also manifested in doing well at school, obtaining a college education and entering a white-collar occupation.

6. ALCOHOLISM

Another manifestation of socioeconomic-status differences in conscientiousness is alcoholism. Excessive alcohol consumption, like smoking, betrays an inability to control impulses in the light of adverse future consequences, and it shows the same inverse socioeconomic-status gradient. Illustrative data for Britain, Finland and the United States derived from social surveys carried out in the 1970s and 1980s is shown in Table 12.4. The criteria adopted for alcoholism differ slightly between the three countries. For Britain it is ''heavy drinkers,'' for Finland ''drunk at least once a month'' and for the United States ''alcohol-

Table 12.4
Socioeconomic-Status Differences in the Prevalence of Alcoholism (Percentages)

Country	Sex	Socio-economic status 1	2	3	4	5	Reference
Britain	Male	9	15	33	32	34	Coleman and Salt,
	Female	0	1	2	3	2	1992
Finland	Male	8	16	25	24	--	Hassan, 1989
	Female	3	2	--	--	--	
United States	Male	6	10	15	--	14	Robins and Regier,
	Female	1	3	1	--	3	1991

Note: 1 = professional; 2 = clerical; 3 = skilled; 4 = semiskilled; 5 = unskilled.

ics,'' but in spite of these differences the inverse association with socioeconomic status is clear and substantial. The inverse gradient is also present between excess alcohol consumption and educational level. For the United States Robins and Regier (1991, p. 101) report a large epidemiological study giving lifetime prevalence rates for alcoholism of 16.4 (high school dropouts) and 10.0 (college graduates).

7. SEXUAL RESTRAINT

Another important area of life in which socioeconomic-status differences in conscientiousness are expressed is sexual restraint. The first extensive data showing greater sexual restraint in the middle class were collected by Alfred Kinsey and his team in the United States in the 1940s and 1950s. The Kinsey data have been analyzed by Rushton and Bogaert (1988) for white college- and noncollege-educated people, and the results indicated that the college educated showed greater sexual restraint in several respects. For instance, 21 percent of the college educated as compared with 40 percent of the noncollege educated had had sexual intercourse by the age of 17, and 30 percent of the college educated as compared with 37 percent of the noncollege educated had extramarital sexual relationships (both these differences are statistically significant).

Similar class differences have been found in Britain. For instance, a survey

Table 12.5
Age of First Sexual Intercourse in Britain by Socioeconomic Status and Age Group

		Socio-economic status					
Sex	Age	1	2	3NM	3M	4	5
Men	16-24	18	17	17	16	16	16
Women	16-24	18	17	17	17	16	16
Men	25-34	18	17	17	16	16	16
Women	25-34	19	18	18	17	17	17
Men	35-44	19	18	18	17	18	17
Women	35-44	19	19	18	18	18	18
Men	45-59	21	19	19	18	18	17
Women	45-59	22	20	20	19	20	20

Source: Wellings et al. (1994).

of 1,831 16 to 45-year-olds found that 15 percent of socioeconomic classes 4 and 5 had sexual intercourse by the age of 16, as compared with 4 percent of classes 1 and 2 (Gorer, 1971). The same difference was found by Schofield (1965). A major British survey of sexual behavior was carried out in 1990 on a representative sample of 18,555 by Wellings, Field, Johnson and Wadsworth (1994). Their results for the age of first sexual intercourse for men and women separately, and for four age groups, are shown in Table 12.5. Notice the general trend for those in social classes 3M, 4 and 5 (skilled, semiskilled and unskilled) to begin their sexual lives two or three years earlier than those in classes 1, 2 and 3NM. The same also emerged in attitude surveys. For instance, Eysenck (1976) has constructed a scale of attitudes toward a variety of sexual behaviors, such as wife swapping, extramarital intercourse and so on, and found middle-class respondents generally more moral than the working class.

Numerous further studies have shown that precocious adolescent sexuality is more prevalent among the lower socioeconomic classes. For instance, Miller and Sneesby (1988) report an investigation of 836 high-school students aged 15 to 18 in Utah and New Mexico. Sexual experience and permissive attitudes correlated −0.10 and −0.15 with fathers' education attainment, and −0.29 and −0.31 with school grades. The authors believe that poor performance in school is responsible for permissive sexual attitudes among adolescents and motivates them toward sexual intercourse, but this is surely an implausible hypothesis. The most probable interpretation of the result is that middle-class children tend to be more conscientious, do better at school and are more sexually restrained.

8. CRIME

A major expression of poor conscientiousness is crime. The criminal displays precisely the syndrome of poorly developed moral sense and lack of restraint, self-control and regard for the well-being of others that is the essence of low conscientiousness. If evidence is needed for the association between low conscientiousness and crime, it can be found in a study carried out in New Zealand by Krueger, Schmutte, Caspi, Moffit, Campbell and Silva (1994). They report a survey of 862 male and female 18-year-olds, which measured the personality trait of *constraint*, a synonym for conscientiousness, and crime. The correlations between strong constraint and low self-reported crime were .44 and with court convictions .15. The results suggest that court convictions are a relatively poor measure of actual crimes committed, probably because relatively few crimes are detected and result in court convictions.

There is overwhelming evidence that crime is much more prevalent among the working class than in the middle class. This class difference in crime has been found in many countries and some 300 studies summarizing the evidence have been reviewed by Braithwaite (1979). The social-class difference is present both in convictions and in crimes admitted in self-report studies, showing that it does not arise simply because the police are more likely to act against working-class individuals.

Some representative figures for socioeconomic-status differences in crime are shown for Britain, Finland and the United States in Table 12.6. The British figures come from the national sample born in 1946 and followed up over their life span, and the figures are percentages convicted of serious offenses and any offenses as teenagers (Douglas, Ross, Hammond and Mulligan, 1966). The Finland data come from Jarvelin, Laara, Rantakallio, Moilanen and Isohanni (1994) who report a survey of 6,007 males born in 1966. Crime rates are for convictions between the ages of 15 and 22. The American data comes from the data of the National Youth Survey of 1,725 11 to 17-year-olds carried out initially in 1977, analyzed by Elliott and Huizinga (1983) and are for self-reported crimes. The data taken as a whole indicate that crime is around three to five times as prevalent among the lower working class as compared with the middle class.

Table 12.6
Socioeconomic-Status Differences in Crime (Percentages)

			Socioeconomic Status			
Country	Crime	Sex	Middle	Upper Working	Lower Working	Reference
Britain	Serious offense	both	2.5	6.1	13.8	Douglas et al., 1966
Britain	Any offense	both	5.5	9.7	18.7	Douglas et al., 1966
Finland	Any offense	both	3.5	7.0	9.0	Jarvelin et al., 1994
United States	Assault	male	2.8	7.6	10.3	Elliott and Huizinga, 1983
United States	Robbery	male	0.8	3.2	5.9	Elliott and Huizinga, 1983
United States	Damage to property	male	16.2	43.5	59.8	Elliott and Huizinga, 1983
United States	Assault	female	1.0	1.4	3.2	Elliott and Huizinga, 1983
United States	Damage Robbery	female	0.3	0.4	1.1	Elliott and Huizinga, 1983
United States	Damage to property	female	6.3	9.3	9.3	Elliott and Huizinga, 1983

9. PSYCHOPATHIC PERSONALITY

We turn finally to socioeconomic-status differences in the prevalence of psychopathic personality. This psychiatric term is used to describe individuals highly deficient in moral sense and the control of antisocial behavior, and whose principal symptoms are a callous indifference to the feelings of others, disregard for the law, aggression, poor work motivation, sexual promiscuity and drug addiction. It is known from self-report studies in which people describe their own behavior that all these expressions of psychopathic personality tend to cluster together (Tygart, 1991) to form a syndrome. This condition has also been called the sociopathic personality. In 1980 the American Psychiatric Association adopted the term antisocial personality disorder and, more recently, borderline personality disorder. All these terms designate the same social pathology, for which I shall retain the term psychopathic personality or psychopath.

A number of studies have demonstrated that psychopaths have a very low level of conscientiousness. For instance, Costa and McCrae (1990) report correlations of −0.40 and −0.42 between conscientiousness and psychopathic personality as measured by questionnaires. Further corroborating studies have been summarized by Harpur, Hart and Hare (1994). In one study on a student sample they report a significant correlation of −0.38 (strong psychopathic personality, weak conscientiousness). In another student sample they report a correlation of −0.49 and in a sample of convicted prisoners a correlation of −0.77. A study by Clark and Livesley (1994) gives a correlation of −0.44 between antisocial behavior and conscientiousness.

The strongest evidence on socioeconomic-status differences in the prevalence of psychopathic personality comes from the American Epidemiologic Catchment Area study on a sample of approximately 20,000 individuals carried out in the early 1980s under the direction of Robins and Regier (1991). They analyzed the prevalence of all the major psychiatric disorders in relation to educational level rather than socioeconomic status, but these are sufficiently closely associated for one to stand as a proxy for the other. Their results are shown in Table 12.7. Notice that for both males and females psychopathic personality is over four times as prevalent among high-school dropouts as among college graduates.

10. CONSCIENTIOUSNESS AND SOCIAL MOBILITY

We have now seen that conscientiousness is strongest in the professional and middle socioeconomic classes and declines progressively among the skilled, semiskilled and unskilled, to reach its nadir among the long-term unemployed.

Why is this? The explanation is that conscientiousness contributes to social mobility, so that in each generation children born with strong conscientiousness tend to rise in the socioeconomic-status hierarchy, while those with weak conscientiousness tend to fall. The principal reason for this is that conscientious adolescents work hard at school and in college, do well in the educational system and obtain the educational credentials which are the necessary passports into professional, managerial and skilled occupations. Once they are on the first rung of the occupational ladder, the conscientious perform well and advance up it.

So far as the contribution of conscientiousness to educational attainment is concerned, one of the first studies was carried out by Wiggins, Blackburn and Hackman (1969). They arranged for students in graduate school to be rated for conscientiousness by their classmates and found that this correlated .50 with their grade point average. The research literature up to the late 1970s was

Table 12.7
Prevalence of Psychopathic Personality in Relation to Educational Level in the United States

Condition	Sex	High School Dropout	High School Graduate	Some College	College Graduate
Psychopathic personality	Male	6.8	3.2	4.3	1.9
Psychopathic personality	Female	1.0	0.3	1.2	0.2

Source: Robins and Regier (1991).

assembled by Jencks (1979) who estimated a multiple correlation between conscientiousness and educational attainment of .56. Subsequent studies confirming this positive association have been published by Hirschberg and Itkin (1978) and by Rothstein, Paunonen, Rush and King (1994). If Jencks's correlation between conscientiousness and educational achievement is accepted as about right, then the contribution of conscientiousness to educational achievement is about the same as that of intelligence, which is generally estimated at around .60 (Eysenck, 1979; Brody 1992).

There is also a substantial body of evidence showing that conscientiousness contributes to occupational status. One of the first studies showing this was published by Elder (1968) who found that the strength of motivation assessed by observers among high-school students correlated .22 with their subsequent occupational achievement. Jencks (1979) analyzed a sample from Kalamazoo in which 389 16-year-old male high-school students were rated by their teachers on various character qualities, and these were examined in relation to their initial occupational status. Positive correlations were found for the following: cooperativeness (.29), dependability (.32), emotional control (.25), industriousness (.34), integrity (.22) and perseverence (.23). All of these traits reflect different aspects of conscientiousness.

Another study showing the contribution of conscientiousness to socioeconomic status has been published by Tomlinson-Keasey and Little (1990). They examined the data file on the Terman sample of approximately 1,500 highly intelligent children followed up over their life span to try to find out why a minority of them failed to achieve the high educational and occupational status which would be expected from their high intelligence. They examined the teachers' and parents' ratings of the gifted sample on 25 personality qualities, which were reduced by factor analysis to three. The first of these they identified as *social responsibility.* The highest correlate of this was conscientiousness and this is essentially the nature of the trait. This factor was found to correlate .87 with educational attainment and .68 with socioeconomic status in midlife. There was a zero correlation between this factor and IQ and mental health. The conclusion to be drawn from this study is that strong conscientiousness has been a major determinant of educational and occupational achievement among this sample and that weak conscientiousness is the major reason for a minority failing to achieve their intellectual potential.

A closer look at the question of why conscientiousness should be associated with occupational status has been taken by Barrick and Mount (1991). On the basis of a meta-analysis of 117 studies in the research literature up to the late 1980s, they find that conscientiousness contributes positively to proficiency in job training and job performance with an overall estimated correlation of .22.

The conclusion to be drawn from this body of work is that conscientiousness contributes about equally with intelligence to both educational and occupational achievement. Through this process individuals with high intelligence and consientiousness tend to move up in the socioeconomic-status hierarchy, while those with low intelligence and conscientiousness tend to fall. Over the course of generations, this process leads inevitably to the stratification of society for both intelligence and conscientiousness.

11. IQ × CONSCIENTIOUSNESS = ACHIEVEMENT

Conscientiousness therefore contributes to social mobility in the same way as intelligence does. We can express the joint operation of the two traits in the form of the equation Intelligence × Conscientiousness = Achievement.

We noted in Chapter 2 that this general formula was first proposed by Francis Galton (1869) in his *Hereditary Genius*, where he used the terms *zeal* and *energy* for what has come to be known as conscientiousness, and that essentially the same formula was proposed by the British sociologist Michael Young in his book *The Rise of the Meritocracy.* Jensen (1980, p. 241) proposed the equation Aptitude × Motivation × Opportunity = Achievement. It is not clear why Jensen prefers the concept of aptitude rather than intelligence. Probably he has in mind an individual's particular constellation of abilities rather than his or her general intelligence. For instance, someone with strong verbal aptitude might succeed as a writer. Strictly speaking, both general intelligence and the relevant aptitudes should be present in the equation because both contribute to achievement in any particular vocation.

Notice that Jensen adds opportunity to his specification equation for achievement. No doubt in past centuries there were many people who had both the necessary intelligence and motivation for success but who lacked the opportunity. How far this is the case in Western nations in the second half of the twentieth century is a moot issue. It is arguable that everyone has the opportunity to obtain educational credentials and secure a reasonably high-status occupation and therefore that differences in opportunities have little effect on achievement.

Others will take the view that significant differences in opportunity are still present and that the Western nations still have some way to go before they become full meritocracies in the sense that social status is determined solely by merit. We need not pursue the issue here.

Notice also that Jensen links his causal variables by multiplication signs rather than the addition signs used by Young. The reason for this is that if either aptitude or motivation are zero, achievement becomes zero in the multiplicative formulation, whereas some achievement is possible if the variables are combined additively. The multiplicative relationship is preferable, because obviously someone with zero aptitude or motivation is going to have zero achievement.

Thus Galton, Young and Jensen have all envisaged achievement as a joint function of intelligence or aptitude, and effort or motivation. A similar formula has been advanced by Cattell (1971, p. 387), who proposed an equation for achievement consisting of intelligence plus a number of personality traits of which the most important was what he calls "super-ego strength," a trait which is essentially the same as conscientiousness. Cattell proposed that the positive contribution of his super-ego strength factor operates through consistent motivation and self-discipline to work for long-term goals, starting with the conscientious performance of homework in adolescence.

My own preferred variant of the equation is IQ $\times$ Conscientiousness $\times$ Opportunity = Achievement. I have replaced the relatively narrow trait of motivation by the broader trait of conscientiousness, which includes work motivation, a strong work ethic and the ability to delay gratification, to work for long-term goals and to resist peer pressures for time-wasting distractions, experimentation with drugs and so on. This formula explains why it is that those with high intelligence and conscientiousness are upwardly socially mobile and have come to be disproportionately represented in the professional and middle classes, while those with low intelligence and conscientiousness are correspondingly downwardly mobile and over-represented in the lower classes.

12. CONCLUSIONS

This chapter has been devoted to the demonstration that there is a socioeconomic-status gradient for the trait the eugenicists called character and that is known in contemporary psychology as conscientiousness. The eugenicists believed that the professional and middle classes are stronger in regard to this trait than the working classes. We have seen evidence from a wide variety of sources that this belief was correct—evidence drawn from social-class differences in the values of professionalism, thrift and the capacity to delay gratification; in moral values and the strength of the work ethic; in personal responsibility regarding the preservation of health by not smoking or drinking alcohol excessively; in sexual restraint, crime and the prevalence of psychopathic personality.

We have seen that the reason for the association between conscientiousness and socioeconomic status lies in the contribution of conscientiousness to social mobility. In each generation children with high conscientiousness tend to rise in the social hierarchy, largely through the positive effect of conscientiousness on educational achievement and the efficient performance of work; while those with low conscientiousness tend to fall. This process leads to a concentration of individuals with high conscientiousness at the top end of the socioeconomic-status hierarchy and of those with low conscientiousness at the bottom.

The eugenicists believed further that conscientiousness, as we now call it, has a genetic basis; and therefore that the relatively low fertility of the professional classes implies genetic deterioration of the population for conscientiousness. Whether the eugenicists were right in their belief that conscientiousness has a genetic basis is the question we examine in the next chapter.

Chapter 13

The Genetic Basis of Socioeconomic-Status Differences in Conscientiousness

1. The Heritability of Conscientiousness. 2. The Work Ethic. 3. Crime. 4. Psychopathic Personality. 5. Alcoholism. 6. Smoking. 7. Genetic Basis for Socioeconomic-Status Differences in Crime. 8. Genetic Determination of Socioeconomic Status. 9. Genetic Processes in the Intergenerational Transmission of Socioeconomic Status. 10. Conclusions.

In this chapter we examine the question of whether conscientiousness has any genetic basis. We look first at heritability studies of questionnaire measures of the trait, and then at the heritability of its expressions in the work ethic, in crime, psychopathic personality, alcoholism and cigarette smoking. We shall see that all of these have substantial heritabilities. We conclude by arguing that the social-class differences in conscientiousness shown in Chapter 12 must have a genetic basis and present direct evidence that social class itself is genetically determined.

1. THE HERITABILITY OF CONSCIENTIOUSNESS

Most of the studies of conscientiousness have used the method of comparing the correlations on the trait for identical and same-sex nonidentical or fraternal twins. The methodology is the same as that used for the estimation of the heritability of intelligence. If the correlation for identical twins is greater than for nonidentical, the trait must have some heritability, and the magnitude of the heritability can be quantified by Falconer's (1960) formula of doubling the difference between the two correlations.

A list of seventeen studies comparing the correlations for identical and nonidentical twins for conscientiousness and related characteristics is set out in Table 13.1. Shown here are the traits investigated; the sex of the twins; the total

Table 13.1
Identical- and Fraternal-Twin Correlations and Heritabilities for Conscientiousness and Related Traits

Trait	Sex	Number of pairs	Correlations		Heritability	Country	Reference
			Identical	Fraternal			
Conscientious-ness	Both	110	.50	.17	.66	United States	Cattell, Schnerger and Klein, 1982
Conscientious-ness	Both	299	.47	.11	.72	Sweden	Bergeman et al., 1993
Psychopathic Deviate	Both	68	.57	.18	.74	United States	Gottesman, 1963
Responsibility	M	354	.57	.27	.60	United States	Loehlin and Nichols, 1976
Responsibility	F	496	.43	.38	.10		Loehlin and Nichols, 1976
Socialization	M	354	.53	.16	.74		Loehlin and Nichols, 1976
Socialization	F	496	.55	.48	.14		Loehlin and Nichols, 1976
Self-control	M	354	.56	.25	.62		Loehlin and Nichols, 1976
Self-control	F	496	.57	.36	.42		Loehlin and Nichols, 1976
Constraint	Both	431	.58	.25	.66	United States	Tellegen et al., 1988

Altruism	Both	477	.53	.25	.56	Britain	Rushton et al., 1986
Empathy	Both	477	.54	.20	.68		Rushton et al., 1986
Nurturance	Both	477	.49	.14	.70		Rushton et al., 1986
Psychoticism	M	117	.53	.16	.74	Britain	Eaves, Eysenck and Martin, 1989
Psychoticism	F	301	.41	.07	.68	Britain	Eaves, Eysenck and Martin, 1989
Callousness	Both	175	.63	.29	.74	United States	Livesley, Jang, Jackson and Vernon, 1993
Empathy	Both	230	.41	.05	.72	United States	Matthews, Batson, Horn and Rosenman, 1981
Psychoticism	M	919	- -	- -	.50	Australia	Martin and Jardine, 1986
Psychoticism	F	1984	- -	- -	.36	Australia	Martin and Jardine, 1986
Antisocial Behavior	M	99	- -	- -	.45	United States	Rowe, 1986
Antisocial Behavior	F	166	- -	-	.65	United States	Rowe, 1986

number of pairs, approximately half of which are identical and half nonidentical; the correlations for identical and nonidentical pairs; the heritability obtained by doubling the difference between the identical and nonidentical correlations representing the heritability coefficient; the country in which the investigation was carried out; and the reference giving the authors of the study. The last four rows of the table present data from studies by Martin and Jardine (1986) and Rowe (1986) which did not give the correlations between identical and nonidentical twins but used an alternative method for calculating the heritabilities.

The crucial feature of the set of studies is that all the heritabilities are positive and most of them are appreciable. The median of the 21 heritabilities is .66, and this can be adopted as the best working estimate of the heritability of conscientiousness. Thus, about two-thirds of the variability in conscientiousness is genetically determined and the remaining one-third is determined by environmental influences.

A criticism that has sometimes been made of the identical-nonidentical twin comparison method is that parents may treat identical twins more similarly and that this could explain any greater similarity in their personality. This possibility has been investigated by Loehlin and Nichols (1976); they found there was no tendency for twins treated alike to be more alike in personality than those not so treated. This objection seems therefore to be invalid.

There have also been studies estimating the heritability of conscientiousness using identical twins reared apart and examining their correlation in the trait. The leading study is by Tellegen, Lykken, Bouchard, Wilcox, Segal and Rich (1988) who examined a sample of 44 such twins who were tested as adults with the Multidimensional Personality Questionnaire (MPQ), a questionnaire measuring fourteen personality traits including *constraint*, a measure of conscientiousness. The correlation for these identical twins for constraint was .57, giving a direct measure of the heritability of the trait of 57 percent. This figure should be corrected for the imperfect reliability of the measure, but it is not clear what reliability coefficient should be used to make this correction. Any reasonable estimate would raise the heritability to about the same figure of .66 obtained from the identical–fraternal twin comparisons.

In addition to these two twin methods there are studies which corroborate the conclusion that conscientiousness has an appreciable heritability. The first of these consists of the comparison of adopted children for this trait with their biological and adoptive parents. If the trait has a heritability, adopted children will show some resemblance to their biological parents. A study of this kind has been carried out by Willerman, Loehlin and Horn (1992). They measured psychopathic tendencies, a measure of low conscientiousness, in 138 adoptees and their biological and adoptive mothers using the psychopathic deviate questionnaire of the Minnesota Multiphasic Personality Inventory. The results showed that the biological mothers' psychopathic deviate score predicted the score of the child approximately three times as strongly as the adoptive mothers' score, indicating that genetic determination of psychopathic tendencies is about three times as strong as environmental determination.

A further method for estimating the genetic contribution to conscientiousness is to examine the degree of similarity for the trait between unrelated individuals reared together in the same adopting families. If the trait is environmentally determined, these biologically unrelated individuals will resemble one another on the trait as a result of family influences. If the trait is largely genetically determined, the degree of resemblance will be small. A major study of this kind has been carried out by Loehlin, Willerman and Horn (1987), and the results are summarized in the first two rows of Table 13.2. Notice that there is no resemblance for conscientiousness between adoptive parents and their adopted children, or between pairs of adopted children. Also shown in the table are the correlations for the related trait of psychoticism, for various relationships between biologically unrelated individuals in adopting families. None of the correlations is statistically significant and taken overall they average out at approximately zero. This shows that the family environment has no effect on the trait, contrary to most child development theories which hold that personality traits are largely formed in the family. The inference to be drawn from these zero correlations between family members in adopting families is that the trait must be determined by genetic factors and by environmental effects on the trait from outside the family. Presumably these effects consist of the influence of friends, teachers and so on. This may seem a counter-intuitive conclusion but it is the one most commonly drawn by those working in this field, such as Plomin (1986) and Rowe (1994).

2. THE WORK ETHIC

We saw in Chapter 12 that a strong work ethic and the related concept of achievement motivation are major components of conscientiousness, embodying as they do the moral commitment to work and to society by making the best use of one's talents. At the low extreme, an absence of any feeling of moral obligation to work is a central feature of those excessively unconscientious individuals known as psychopathic personalities. For this reason we look now at studies of the heritability of the work ethic.

These studies have used a variety of questionnaire measures of the work ethic, variously designated achievement, achievement via conformity, and achievement via independence. Both the methods of identical twin–nonidentical twin comparisons and of identical twins reared apart have been used in studies of this kind. The results of five sets of data for identical–nonidentical twin correlations and the heritabilities derived from them are summarized in Table 13.3 All five heritability coefficients are appreciable and the average is .52.

In addition there have been two studies of the correlation on this trait of identical twins reared apart. Tellegen, Lykken, Bouchard, Wilcox, Segal and Rich (1988) report results on 44 pairs of such twins given a questionnaire of

Table 13.2
Correlations for Conscientiousness and Psychoticism between Biologically Unrelated Individuals Reared in the Same Family

Trait	Unrelated pairs	N. Pairs	Correlation	Reference
Conscientious-ness	Adoptive parent : adopted child	358	.01	Loehlin, Willerman and Horn, 1987
Conscientious-ness	Adopted child : adopted child	118	-.09	Loehlin, Willerman and Horn, 1987
Psychoticism	Adoptive father : adopted son	18	-.39	Eaves, Eysenck and Martin, 1989
Psychoticism	Adoptive father : adopted daughter	75	.10	Eaves, Eysenck and Martin, 1989
Psychoticism	Adoptive mother : adopted son	26	.05	Eaves, Eysenck and Martin, 1989
Psychoticism	Adoptive mother : adopted daughter	101	-.02	Eaves, Eysenck and Martin, 1989
Psychoticism	Adopted female : adopted female	24	.27	Eaves, Eysenck and Martin, 1989
Psychoticism	Adopted male : adopted male	34	-.17	Eaves, Eysenck and Martin, 1989

Table 13.3
Studies of Identical- and Nonidentical-Twin Correlations and Heritabilities on Work Ethic and Achievement Motivation Questionnaires

Trait	N. Pairs	Sex	Identical	Fraternal	Heritability	Reference
Achievemnent via Conformity	354	M	.48	.05	.86	Loehlin and Nichols,1976
Achievement via Conformity	496	F	.44	.26	.36	Loehlin and Nichols,1976
Achievement via Independence	354	M	.57	.39	.36	Loehlin and Nichols,1976
Achievement via Independence	496	F	.39	.42	.24	Loehlin and Nichols,1976
Achievement	311	Both	.51	.13	.76	Tellegen et al., 1988

the value attached to achievement, for which the correlation was .36. This is a direct measure of heritability but requires correction for reliability. The authors do not give an estimate of reliability, but probably .70 would be a reasonable figure, giving a heritability of .51. The same group of researchers has also reported a study of 23 pairs of identical twins reared apart and given a work motivation questionnaire measuring various attitudes to work, including the value attached to achievement and the utilization of one's ability. The correlation between the twin pairs was .43 (Keller, Bouchard, Arvey, Segal and Davis, 1992). On this occasion, the authors correct the correlation for reliability and temporal stability and conclude that this trait has a heritability of .68. Thus, all these studies of the heritability of the work ethic and achievement motivation show that this trait has a significant heritable component, for which the best estimate derived as the median of the seven readings is .51.

3. CRIME

Crimes are committed by those with weak conscientiousness, and the conclusion that conscientiousness is significantly determined by genetic factors receives further confirmation from studies of the heritability of crime. The first study to use the method of comparing the degree of similarity of identical and nonidentical twins to assess a possible genetic contribution to crime was carried out in Denmark in the late 1920s by Johannes Lange (1929). He found that identical twins were much more alike with regard to criminal convictions. He expressed the degree of twin similarity as a concordance rate, which was the percentage of pairs in which the twins were similar in having criminal convictions, and found that this was the case with 77 percent of identical twins and 12 percent of nonidenticals.

A number of subsequent studies have confirmed this result. Seven further investigations have been summarized by Eysenck and Gudjonsson (1989), all of which found greater concordance among identical as compared with nonidentical twins, the average percentage of the whole set of studies being 67 and 30 percent, respectively. An updated review of the literature based on twin studies carried out in North America, Japan, Norway, Germany and Denmark has been published by Gottesman and Goldsmith (1995). For a total of 545 pairs, the concordance rates were 51.5 percent for identicals and 23.1 percent for fraternals.

These studies show that genetic factors play a significant role in crime, but they do not quantify the heritability. The best single study from which to do this consists of the data on 1,913 twin pairs in Denmark analyzed by Cloninger, Christiansen, Reich and Gottesman (1978). They calculated the correlations for criminal behavior for identical males as .70; identical females, .72; nonidentical males, .36; nonidentical females, .43. The heritabilities obtained by doubling the identical-nonidentical differences between the correlations work out at 68 percent for males and 58 percent for females.

These twin comparisons pointing to a significant heritability of crime have been corroborated by a further set of studies of adopted children to determine whether they show any resemblance to their biological parents in respect of crime. If this is found, it can be inferred that some genetic transmission of the criminal behavior is involved.

An American study employing this methodology was carried out by Crowe (1972, 1975). He began by finding 41 female criminal prisoners who among them had had 52 babies and given them up for adoption. At the time of the study these adopted children were between the ages of 15 and 45. Their criminal behavior was assessed and compared with that in a matched control group of adoptees whose biological mothers had no criminal records. The comparison showed that 17 percent of the offspring of the criminal mothers had been imprisoned as compared with none of those of the noncriminal mothers.

Similar results were obtained in a more extensive study carried out in Denmark. Hutchings and Mednick (1977) obtained a sample of 662 adopted men. Among those whose biological fathers had a criminal record but not their adoptive fathers, 22 percent of the sample had criminal records. Among those with noncriminal biological and adoptive fathers, only 10 percent had any criminal conviction. The result indicates that genetic factors are about twice as potent as environmental factors for criminal behaviors. These adoption studies confirm the twin studies in indicating a substantial heritability for crime.

4. PSYCHOPATHIC PERSONALITY

Like crime, psychopathic personality is a manifestation of low conscientiousness, and if conscientiousness has a significant heritability, we should expect that this would also be found for psychopathic personality. A number of studies have shown that this is so. Eight investigations of identical-twin and fraternal-twin correlations for psychopathic personality, showing greater similarity between identical-twin pairs, have been reviewed by Mason and Frick (1994). Doubling the difference between the correlations gives an average heritability of .41.

Further evidence for a significant heritability of psychopathic personality comes from studies of the similarity of adopted children to their biological parents. This first study of this kind was carried out in Denmark by Schulsinger (1972). He obtained 57 psychopathic adoptees and a further 57 nonpsychopathic adopted controls matched for sex, age, social class and other characteristics. He then assessed the biological parents of the two groups for psychopathic personality. His conclusion was that the incidence of psychopathy among the biological parents of the psychopathic adoptees was two and a half times as great as among the parents of the nonpsychopathic controls. This parent-child similarity for psychopathic personality among adopted children indicates a significant genetic basis for the disorder.

A further study of adopted children with psychopathic personality has been carried out in the United States by Cadoret, O'Gorman, Troughton and Heywood (1985). They found 127 male and 87 female adoptees diagnosed as having antisocial personality disorder. Among the males, 40 percent of the biological fathers or mothers also had antisocial personality disorder as compared with 14 percent among the adoptive parents, indicating that the genetic relationship is approximately three times as strong as the environmental relationship. Among the females 19 percent had an antisocial biological parent and 19 percent an antisocial adoptive parent, indicating that genetic and environmental factors contribute equally to antisocial personality disorder in females.

5. ALCOHOLISM

The next index in terms of which to consider the heritability of conscientiousness is alcoholism, an expression of low conscientiousness because it betrays an inability to control immediate gratification and a deficiency of moral control over socially unacceptable behavior. Alcoholism is typically part of a syndrome of psychopathic and criminal behaviors (Bohman, Cloninger, Sigvardsson and von Knorring, 1987). The same is true of other forms of drug addiction, but alcoholism has been the most investigated.

The research evidence indicates that alcoholism runs from generation to generation in certain families and that part of the reason for this lies in genetic transmission from parents to children. There are strong family resemblances for alcoholism. About 25 percent of the male relatives of alcoholics are themselves alcoholics as compared with about 5 percent in the general population (Cotton, 1979). There are also parent-child resemblances for normal drinking. In an Australian study of 507 14 to 16-year-olds carried out by Webster, Hunter and Keats (1994) a correlation of .32 was found between the amount of alcohol consumed by these teenagers and their parents.

The genetic basis of the family similarities for alcoholism has been shown by both twins and adoption studies. One of the leading twin studies has been carried out in Sweden by Pedersen, Friberg, Floderus-Myrhed, McClean and Plomin (1984). They found that the correlations for alcohol consumption among identical and nonidentical twins were .64 and .27, respectively, indicating a heritability of .74. The correlation for identical twins reared apart was closely similar at .71, a direct measure of heritability.

This conclusion has been confirmed in an Australian study by Heath, Meyer, Jardine and Martin (1991). They obtained 3,810 adult twin pairs and measured both the frequency and the quantity of alcohol consumed. The correlations for male and female identical and same-sex fraternal twins and the heritabilities are shown in Table 13.4, the combined average heritability being .47. A study of alcohol consumption by adopted children in Sweden by Cloninger, Bohman and Sigvardsson (1981) points to the same conclusion. They found that 22 percent of adopted sons of alcoholic biological fathers were themselves alcoholics, as compared with a prevalence of 5 percent in the general population.

All these studies from a variety of countries and using different methodologies show a substantial heritability of alcoholism. They provide further evidence for the heritability of conscientiousness, the weakness of which is expressed in alcoholism.

6. SMOKING

The final index of conscientiousness to be considered is cigarette smoking which is, like alcohol consumption, an expression of weak self-control over immediate impulse gratification. The first people to show that smoking cigarettes has a genetic basis were Todd and Mason (1959). They obtained 60 pairs of identical and same-sex nonidentical twins and found that 83 percent of the identical were concordant for smoking as compared with only 53 percent of the nonidentical.

Several later studies have confirmed that smoking has a significant heritability. In the United States Carmelli, Swan, Robinette and Fabsitz (1992) estimate the heritability at .36 and similar estimates have been obtained from studies carried out in The Netherlands (Boomsma, Koopmans, Van Doormen and Orlebeke, 1994) and in Australia (Hopper, White, Macaskill, Hill and Clifford, 1992; Heath and Martin, 1993). Two studies have shown that both smoking and alcohol consumption are, to some degree, determined by the same genes (Swan,

Table 13.4
Correlations for Identical and Fraternal Twins and Heritabilities for Alcohol Consumption

Sex	Drinking	N. pairs	Identical	Fraternal	Heritability
Males	Frequency	919	.74	.52	.44
Females	Frequency	1984	.66	.32	.68
Males	Quantity	919	.58	.43	.30
Females	Quantity	1984	.56	.32	.48

Source: Heath, Meyer, Jardine and Martin (1991).

Cardon and Carmelli, 1994; Koopmans and Boomsma, 1993). These may be the genes for the inability to control instant gratification in the light of known adverse future consequences, or for lack of regard for one's health and physical well-being and for the moral disapproval of others, which is frequently incurred by smokers and alcoholics.

7. GENETIC BASIS FOR SOCIOECONOMIC-STATUS DIFFERENCES IN CRIME

The thesis argued in the preceding chapter and continuing in the present is that the eugenicists were right in their contention that there are genetically based differences among the socioeconomic classes for what they called character, and what in contemporary psychology is generally termed conscientiousness. We have now reviewed the two major strands of evidence for this view, namely, that socioeconomic-status differences in conscientiousness exist—the evidence for which was set out in Chapter 12—and that conscientiousness has a substantial heritability—the evidence for which has been laid out in this chapter. We now consider direct evidence showing that socioeconomic-status differences in crime are genetically determined.

Van Dusen, Mednick, Gabrielli and Hutings (1983) obtained 14,427 adopted children in Denmark. They examined the criminal records of these when they were teenagers in relation to the socioeconomic status of their biological and adoptive parents. The results showed that children born of high socioeconomic-status parents had lower crime rates than those of middle socioeconomic-status parents, while those born to parents of low socioeconomic status had the greatest crime rates. These results are shown for male and female adoptees separately in Table 13.5 and give the percentage of the adoptees with criminal convictions analyzed by the social class of their biological and adoptive parents. Notice that among adopted males, 16.0 percent of those with low-class biological parents had criminal convictions as compared with 11.6 percent of those with high-class

Table 13.5
Criminal Convictions (Percentages) in Relation to the Social Class of Biological and Adoptive Parents in Denmark

Parents' Social class	Biological Parents Males	Biological Parents Females	Adoptive Parents Males	Adoptive Parents Females
High	11.6	1.0	11.6	2.0
Middle	14.3	2.6	15.6	2.4
Low	16.0	3.0	17.2	3.2

Source: Van Dusen et al. (1983).

parents; among females the disparity is greater at 3 percent as against 1 percent. There is only one possible explanation for this result: the high-class parents must have had fewer genes predisposing their children to commit crimes.

Notice that in these results, the crime rates of the adoptive children are also related to the social class of their adoptive parents to about the same extent as to the social class of their biological parents. This shows that social class–related socialization practices and peer group influences determine crime to about the same extent as genetic factors. This confirms the general belief that conscientiousness is developed through environmental factors such as child socialization, as well as being significantly under genetic control. It should be noted, however, that in normal families the more effective socialization practices of middle-class parents are not a solely environmental influence. They arise partly because middle-class parents tend to be genetically more conscientious and consequently exert stronger socialization pressures on their children.

8. GENETIC DETERMINATION OF SOCIOECONOMIC STATUS

We have now seen that all the major components and expressions of conscientiousness have substantial heritabilities. We saw in Chapter 12 that there are differences in conscientiousness between the socioeconomic classes, such that conscientiousness is strongest in the professional and middle classes, and that these social-class differences are reinforced in each generation through the process of social mobility, by which people reallocate themselves in each generation to the social class appropriate for their level of conscientiousness. Whatever their social origin, through social mobility, children born with the genes for strong conscientiousness tend to rise in the social hierarchy and end up in the professional and middle classes, while those with genes for weak conscientiousness tend to fall into the lower classes. The result of this is that the socioeconomic classes have to some extent become genetically differentiated for conscientiousness, just as they are for intelligence, as explained in Chapter 11.

Table 13.6
Correlations for Identical and Nonidentical Male American Twins and Heritabilities

	Identical	Non-identical	Difference	Heritability
Schooling	.76	.54	.22	.44
Initial occupation	.53	.33	.20	.40
Adult occupation	.60	.43	.23	.46
Income	.55	.30	.25	.50

Source: Taubman (1976).

In the first half of the twentieth century, eugenicists understood that the process of social mobility inevitably leads to some degree of differential segregation of the genes for conscientiousness, as well as those for intelligence, into the social classes. R.A. Fisher wrote on this issue that "the social classes become genetically differentiated, like local varieties of a species, by the agencies controlling social promotion or demotion" (1929, p. 245). Fisher was well aware that what he called "the agencies controlling social promotion or demotion" included the personality traits of conscientiousness, in addition to intelligence.

In Chapter 11 we noted Herrnstein's syllogism proving the genetic basis of social-class differences in intelligence. The syllogism can be extended to social-class differences in conscientiousness and can be stated formally as follows: (1) if conscientiousness has some heritability and (2) if conscientiousness makes a significant contribution to socioeconomic status, then (3) social mobility will bring about genetically based differences between the social classes in conscientiousness.

The logic of Herrnstein's syllogism is unassailable, but all syllogisms gain additional credibility if their conclusion can be demonstrated empirically. So we look now at direct evidence that socioeconomic status is to some degree genetically determined. The first study to demonstrate this was carried out by Taubman (1976). He obtained 1,019 pairs of male identical twins and 907 pairs of male nonidentical twins with an average age of 51. The twins were white American World War II veterans who were mailed a questionnaire in the early 1970s asking for details of schooling, first occupation, present occupation and incomes. For all four variables the identical twins were considerably more similar than the fraternal, indicating that genetic factors are involved. The correlations for the identical and nonidentical twins, and the heritabilities calculated by doubling the difference between the correlations for the two types of twins, are shown in Table 13.6. We see here that schooling, occupational status and income all have a substantial genetic component with heritabilities between 40 and 50 percent.

The next studies to corroborate this conclusion were a series of investigations carried out in Denmark by Teasdale and his colleagues. In the first of these, information was obtained on the occupations of 13,194 adopted individuals and on the occupations of their biological and adoptive parents (Teasdale, 1979). The occupations were scored for socioeconomic status, and the status of the adopted children was found to be correlated .18 with that of their biological parents and .21 with that of their adoptive parents. The positive correlation of the socioeconomic status of adopted children with that of their biological parents indicates the genetic transmission of socioeconomic status. The correlations are low but this is partly because the scale for measuring socioeconomic status is "a blunt instrument," as Teasdale aptly describes it.

In the second study Teasdale and Owen (1981) examined the Danish adoption sample and extracted 2,948 individuals who had siblings who had also been adopted by different families. The socioeconomic status of the adopted children was obtained when they were at an average age of 39. There was a significant correlation of .22 between the socioeconomic status achieved by pairs of separated siblings reared in different adoptive families. This can only be due to the common inheritance of the pairs of siblings, indicating some genetic determination of socioeconomic status. This study also found a positive correlation of .17 for the socioeconomic status of unrelated pairs of individuals adopted in the same family, showing an environmental determination of socioeconomic status. Notice that in this study the correlation for biological siblings reared apart was higher than for unrelated individuals reared together, indicating that genetic factors are more important than environmental as determinants of socioeconomic status.

The third study in this series was carried out by Teasdale and Sorensen (1983) on data for the socioeconomic status of 1,417 adopted adults obtained at an average age of 39, and for the socioeconomic status of their adoptive and biological fathers. The correlations for the socioeconomic status of the subjects were .25 with the socioeconomic status of the adoptive father and .20 with the socioeconomic status of the biological father in the case of the adopted men, and .18 and .11, respectively, in the case of the adopted women. The positive and statistically significant correlations between the socioeconomic status of adopted children with those of their biological fathers can only mean that the traits determining socioeconomic status must be, to some degree, inherited. In this study the adopted children's socioeconomic status was associated a little more strongly with that of their adoptive parents than with that of their biological parents.

In the fourth study Teasdale and Owen (1984a) obtained data on approximately 29,000 Danish males born in the years 1944–1947, of whom 273 had been adopted by unrelated individuals. Information was also obtained for the social class of their biological fathers, their adoptive fathers and the men's social class at approximately 36 years of age. The data yield three correlations, all of which are statistically significant. The first is for the social class of biological fathers and their biological sons and is .37, indicating that sons significantly resemble their fathers in regard to social class. The correlation is identical to that arrived at by Jencks (1972) from the leading American studies. The next correlation is for adoptive fathers and adopted sons and is .17. This is less than half the correlation between biological fathers and their sons. Why this difference? It can only be because the adopted sons do not inherit the genes of their adoptive fathers. The third correlation is between biological fathers and adopted sons and is .23. Why is there a statistically significant correlation between the social class of biological fathers and their sons who were reared in adoptive families, never having seen their fathers? The answer can only be that the genes determining social class have been transmitted from fathers to sons.

The final study in this series obtained data on a further 88 pairs of biological siblings adopted and reared in different families (Teasdale, Sorensen and Owen, 1984). The siblings' socioeconomic status was ascertained at the age of 40 and the correlation for the pairs was .24. Thus, pairs of siblings have some similarity for socioeconomic status even when they have been reared in different environments. The only explanation can be that they inherit qualities that determine their socioeconomic status. Teasdale's studies examined the possible effects of selective placement and found that they did not affect the positive biological father–adopted child associations.

Another study showing that social class has a genetic basis comes from Norway. Tambs, Sundet, Magnus and Berg (1989) obtained 570 pairs of identical and 575 pairs of nonidentical male twins and found that the identicals were more similar in respect of their adult socioeconomic status than the fraternals. They do not give the correlations but use an alternative method to calculate the heritability of socioeconomic status at 41 percent.

The most recent study on this issue comes from Sweden. Lichtenstein, Herschberger and Pedersen (1995) report on 24 male identical twins separated shortly after birth and brought up in different families. The correlation for their adult socioeconomic status was .50, indicating a minimum of 50 percent heritability. They also report on 42 identical and 50 nonidentical male twins reared together. The correlations for adult socioeconomic status were .84 and .36, respectively. Doubling the difference between the two correlations gives a heritability of 96 percent.

The reason that socioeconomic status has a substantial heritability is that the factors that determine socioeconomic status have high heritabilities. We saw in Chapter 12 that these are intelligence and conscientiousness. The substantial heritabilities of socioeconomic status obtained from the studies carried out in the United States, Denmark, Norway and Sweden provide a direct demonstration that the social classes differ genetically.

9. GENETIC PROCESSES IN THE INTERGENERATIONAL TRANSMISSION OF SOCIOECONOMIC STATUS

The genetic processes involved in the intergenerational transmission of socioeconomic status are similar, although more complex, compared to those involved in the intergenerational transmission of intelligence. The additional complexity arises from the joint operation of intelligence and conscientiousness in the determination of socioeconomic status in accordance with the general formula IQ × Conscientiousness = Achievement. It follows from this formula that the genes for both intelligence and conscientiousness have come to be concentrated most strongly in the professional class, and progressively less strongly in other classes down the socioeconomic-status hierarchy. The result of this is that the social classes tend to breed true because children inherit the genes of their parents. This explains the existence of social-class continuities of about 66 percent in societies stratified into three classes: about two-thirds of children remain in the same social class as their parents, as shown for a number of Western nations by Heath (1981).

10. CONCLUSIONS

In this chapter we have seen that conscientiousness has a substantial heritability, whether it is measured directly by questionnaires or indirectly through its expression in the work ethic, crime, psychopathic personality, alcoholism and cigarette smoking. The best estimate of the heritability of conscientiousness obtained from questionnaire measures of the trait derived from the comparison of identical and nonidentical twins, and from the correlation between identical twins reared apart, is .66. All of the five indirect measures of conscientiousness also have substantial heritabilities.

We saw in Chapter 12 that there is a positive association between conscientiousness and socioeconomic status which is reinforced in each generation through social mobility. We have now seen that conscientiousness has a high heritability and inferred that this must mean that the genes responsible for high conscientiousness are disproportionately represented in the professional and middle classes. The social-class differences in conscientiousness cannot only be in the environmental component of the trait. These differences must exist also in the trait's genetic component. There is direct evidence for this conclusion in the form of adoption studies showing that the high prevalence of crime among the lower socioeconomic classes has a genetic basis and from studies showing that socioeconomic status itself is, to a substantial extent, genetically determined.

The eugenicists believed that the professional and middle classes are stronger on conscientiousness than the lower classes, and that this social-class difference has a genetic basis. Once again, the research evidence has proved them right.

Chapter 14

Dysgenic Fertility for Conscientiousness

1. Educational Level, Socioeconomic Status and Fertility Revisited. 2. Conscientiousness, Crime and Number of Siblings. 3. The Fertility of Criminals. 4. The Secular Increase of Crime. 5. Conclusions.

The eugenicists were concerned that natural selection had broken down for the personality trait they called character, and that those deficient in character had begun to have high fertility unchecked by high mortality. They believed that character has a genetic basis, and therefore that the high fertility of those with weak character entailed genetic deterioration for this trait.

We have examined in Chapter 12 the nature of the eugenicists' concept of character, that it has reappeared in contemporary psychology as the trait of conscientiousness, and we examined its expressions in the work ethic and the manifestations of its weakness in crime, psychopathic personality, alcoholism and cigarette smoking. We have seen in Chapter 13 that conscientiousness and its expressions and manifestations have high heritabilities. We are now ready to turn to the evidence on whether fertility for conscientiousness has been dysgenic, as the eugenicists feared.

1. EDUCATIONAL LEVEL, SOCIOECONOMIC STATUS AND FERTILITY REVISITED

We have seen extensive evidence for an inverse association between educational level and fertility in modern populations in Chapter 9, and for an inverse association between socioeconomic status and fertility in Chapter 10. We have also seen that educational level and socioeconomic status are determined by intelligence and conscientiousness, in accordance with the equation Intelligence × Conscientiousness = Achievement. This means that the dysgenic fertility for educational and occupational achievement, the extensive evidence for which has been set out in Chapters 9 and 10, must reflect dysgenic fertility in the underlying components of educational and occupational achievement, namely, intelligence and conscientiousness.

In order to show this formally, we can extend Herrnstein's syllogism—which he originally proposed to demonstrate that in any society in which intelligence has a significant heritability and in which intelligence is a determinant of social status—the social classes will, to some degree, be genetically differentiated for intelligence. In Chapter 13 we applied the syllogism to conscientiousness and argued that the social classes have become genetically differentiated for conscientiousness in the same way. We can now add a corollary to the syllogism;

1. The genes for conscientiousness are disproportionately represented in the higher socioeconomic classes, the evidence for which has been set out in Chapter 13.
2. There is an inverse relationship between socioeconomic status and fertility, the evidence for which has been reviewed in Chapter 10.
3. Therefore, the genetic quality of the population in regard to conscientiousness must be deteriorating.

The same syllogism holds for educational level and can be stated formally as follows:

1. The genes for conscientiousness are disproportionately represented among the better educated, as we argued in Chapter 13.
2. There is an inverse relation between educational level and fertility, as shown in Chapter 9.
3. Therefore, the genetic quality of the population in regard to conscientiousness must be deteriorating.

Thus, both chains of reasoning lead to the same conclusion, that there is dysgenic fertility for conscientiousness in modern populations. Nevertheless, it would be useful to consider whether this conclusion can be corroborated by other kinds of evidence and this is our next task.

2. CONSCIENTIOUSNESS, CRIME AND NUMBER OF SIBLINGS

We saw in Chapter 5 that the first studies made by the eugenicists on the issue of whether there was dysgenic fertility for intelligence consisted of investigations of the relationship between people's intelligence and their number of siblings. Numerous studies showed that the higher people's intelligence, the fewer their siblings, and eugenicists inferred this must mean that intelligent people were having fewer children than the less intelligent. This, they argued, must mean that the genetic quality of the population in regard to intelligence is in decline. We concluded in our account of this work that this reasoning was correct.

The same argument can be used for conscientiousness, and we do so here by taking psychopaths and criminals as the criterion groups of those excessively low on the trait. So the question is, do psychopaths and criminals have more siblings than the more conscientious remainder of the population? The answer to this question has been provided by Ellis (1988) in a review of 32 studies carried out in the United States and Britain. The results are that all except one of these studies found that psychopaths and criminals have more siblings than the general population. Typical of these studies, although not included in Ellis' review, is the investigation carried out by Rutter, Tizard and Whitmore (1981) on 2,334 children in the Isle of Wight in England. Antisocial behavior was assessed by a teachers' rating scale and, on the basis of this, 63 children were designated as having antisocial personality. The results presented by the authors were that 3.2 percent of these came from one-child families and 44.4 percent from families with four or more children. Among a comparison group of socialized children, 11.3 percent came from one-child families and 34.0 percent from families with four or more children. It is clear that children with antisocial personality tend to come from large families, although unhappily it is not possible to estimate the magnitude of the relationship or the extent of dysgenic fertility from the data presented. The authors could have done this, but they did not. At the time of this study, the possibility that fertility for psychopathic personality might be dysgenic had become a taboo subject.

A study which gives some indication of the magnitude of dysgenic fertility with regard to criminal behavior, published after Ellis' compilation, has been reported by Tygart (1991). He obtained 400 male and 400 female 16-year-olds in California and gave them a self-report delinquency questionnaire in which the subjects checked on a 0–4 scale the frequency with which they had committed the criminal acts of breaking and entering, theft, robbery, assault, vandalism, car theft and drug use. Information on number of siblings was also recorded. The correlations between number of criminal acts and number of siblings was 0.41 for males and 0.49 for females. These correlations are about twice as high as the negative correlations between intelligence and number of siblings found in the 1920s–1940s, suggesting that dysgenic fertility for conscientiousness is about twice as large as that for intelligence.

We noted in our discussion of the early work on the negative association between number of siblings and intelligence that there are two problems in inferring the existence of dysgenic fertility from data of this kind. The first is that the childless are excluded from the samples. If criminals, psychopaths and the unconscientious tend to be disproportionately childless, the dysgenic effect arising from the inverse association between sibling size and conscientiousness would be reduced and perhaps eliminated. It seems unlikely that these groups tend to be childless or have lower than average fertility, so this objection can probably be discounted.

More serious is the second problem that the inverse association between conscientiousness and number of siblings might arise solely through environmental processes. This is the usual interpretation advanced by sociologists and criminologists who have noted this association and have generally interpreted it in terms of what is called parental control theory (Tygart, 1991). The theory is that parents cannot control and socialize their children when they have large numbers of them as effectively as they can when they have only one or two, so children from large families tend to turn out poorly socialized, psychopathic, criminal and with low conscientiousness. This is an interpretation similar to that often advanced to explain the inverse association between number of siblings and intelligence, for which it is also proposed that intelligence is depressed by membership in large families.

It is not easy to discount parental control theory. One way of testing it would be to see whether the inverse relationship between number of siblings and crime holds for one and two-child families. If it does not, the parental control theory would be seriously weakened and genetic transmission theory correspondingly strengthened. Another test would be to examine whether the inverse relationship holds in families of adopted children. Again, if it does not, parental control theory would be discounted.

A general argument against parental control theory is that recent work in developmental psychology has led to the conclusion that family effects on personality are quite small (Plomin and Daniels, 1987; Rowe, 1994). One of the principal reasons for this view is that siblings are fairly dissimilar with regard to personality traits, and the extent to which they are similar can be explained by their genetic similarity. Probably parental control theory cannot provide the whole explanation for the association between number of siblings and crime, psychopathic personality and low conscientiousness. Criminals, psychopaths and the unconscientious are precisely the people likely to have larger than average number of children because of their lack of sexual restraint, their inability to plan ahead, their general irresponsibility and their failure to use contraception conscientiously, consistently and efficiently. Probably these people do have larger than average number of children, and this is the major reason why criminals, psychopaths and the unconscientious have large number of siblings.

Nevertheless, as with the interpretation of the negative association between intelligence and number of siblings, the inference of dysgenic fertility from the relationship between crime and large number of siblings is indirect and open to challenge by determined critics. More satisfactory would be a direct demonstration of a negative association between conscientiousness among adults and their completed fertility.

3. THE FERTILITY OF CRIMINALS

One of the most straightforward ways of examining whether there is dysgenic fertility for conscientiousness would be to obtain data on the number of children of criminals taken as a criterion group of the unconscientious, and compare them with the number of children of noncriminals. A result showing that criminals have greater than average number of children would point to the presence of dysgenic fertility for conscientiousness.

When it is considered that the increase of crime has been a major social problem in many Western nations in the second half of the twentieth century, and it is so well established that crime is significantly genetically determined, one would expect that criminologists would have investigated the possible contribution of high fertility among criminals to the secular increase in crime. Curiously, no criminologist appears ever to have asked this question, and even the evidence for the heritability of crime has scarcely penetrated the community of academic criminologists. Thus, in 1994 Maguire, Morgan and Reiner produced *The Oxford Handbook of Criminology*, a volume of over 1,200 pages and claimed to be "the most comprehensive and detailed textbook about criminology ever published," yet no mention whatever was made of the genetic contribution to crime, let alone of the possibility that some of the increases in crime might be attributable to the high fertility of criminals. The editors and contributors to this text are locked into the time warp of solely environmental explanations for crime.

When I began to consider the issue of the fertility of criminals, I had expected that a search through the literature would provide the data I needed. My efforts, however, proved unavailing, and eventually I had recourse to the data for a sample of criminals collected by the Institute of Criminology at the University of Cambridge. The Cambridge cohort consists of a longitudinal study of a sample of 411 boys first selected in 1961 and 1962 at the ages of 8 and 9 as a representative sample from working-class districts in London. The sample has been followed up over subsequent years, during which the focus of the study has been on the characteristics differentiating the boys who developed into delinquents from those who did not. It has been found that the boys who later became delinquent tended to be characterized by low IQs and large family size, and to come from families where the parents had criminal convictions and low incomes. The principal findings of the study up to the mid-1980s have been reported by Farrington and West (1990).

For the purpose of the present inquiry, information has been abstracted from this database for the criminal convictions of the parents of the boys in the sample up to the time of the boys' tenth birthday, and on the number of children of these parents. The parents were divided into two groups consisting of those where one or both had a criminal conviction for an indictable offense, and those where neither of the parents had a conviction for an indictable offense. The average number of children of the parents with criminal convictions ($N = 104$) was 3.91 ($sd = 1.77$), and the average number of children of parents without criminal convictions ($N = 307$) was 3.12 ($sd = 1.46$). The difference between the two groups is statistically significant ($t = 4.16$, $p < .001$). I have reported the study in more detail in Lynn (1995).

The result shows that in this sample the fertility of criminals is higher than that of noncriminals matched for social class and urban residence in London. It is also useful to consider whether the fertility of this sample of criminals is greater than that of the British population as a whole. Average fertilities for British married couples for the marriage cohorts of 1941 to 1945, 1946 to 1950 and 1951 to 1955 (the three relevant comparison groups obtained from census data) were 2.14, 2.19 and 2.29 respectively, averaging 2.21 (Coleman and Salt, 1992, p. 162). Thus, the fertility of 3.91 of this sample of criminals is 77 percent higher than that of the British population as a whole. The results show that this sample of criminals has considerably higher fertility than that of the British population and provides the first direct evidence that fertility for conscientiousness in the late twentieth century is dysgenic.

4. THE SECULAR INCREASE OF CRIME

In Chapter 8 we considered the paradox that although fertility in relation to intelligence has been dysgenic and genotypic intelligence is therefore declining, many studies have shown that phenotypic or measured intelligence has actually been increasing in economically developed nations during most of the twentieth century. A number of people, such as Penrose (1948, 1950) and Duncan (1952), found this difficult to understand and thought there must be some flaw in the argument that the inverse relationship between intelligence and fertility implied a decline in genotypic intelligence. We saw that there is not really any problem here. The answer is simply that environmental improvements, principally in nutrition, have counteracted the genotypic deterioration to such an extent that phenotypic intelligence has risen.

There is no parallel paradox in the case of dysgenic fertility for crime and secular trends in the crime rate. The straightforward prediction is that the high fertility of criminals has led to an increase in the number of genes in the population responsible for crime and this will show up in increasing crime rates. These increasing crime rates have certainly occurred in most of the economically developed nations during the second half of the twentieth century. In the United States, crime rates approximately tripled between 1960 and 1990; in Britain they quadrupled, and similar increases have occurred in many other countries (Wilson and Herrnstein, 1985; Maguire, 1994; Himmelfarb, 1995).

In the case of Britain, it is possible to use the data on the fertility of criminals reviewed in the last section to estimate how far the secular increase in crime is attributable to the high fertility of criminals. It was estimated that criminals have 77 percent more children than noncriminals. If criminal behavior among males has a heritability of .68, as proposed in Chapter 13, crime over one generation should increase by 52 percent ($77 \times .68 = 52$)). The actual increase in crime in Britain over the relevant time period 1950–1975 is approximately 300 percent, from around 500,000 to 2 million crimes a year (Maguire, 1994). This increase is considerably greater than the rise that would be predicted from the greater fertility of criminals, and it suggests that about 17 percent of the secular increase in crime is attributable to the increase in the genes underlying criminal behavior. Hence, the large increase in crime is principally caused by the weakening of social controls on antisocial behavior and other environmental influences. Nevertheless, a 52 percent increase in crime per generation resulting from genetic causes is a matter of concern.

5. CONCLUSIONS

In this chapter we have examined three lines of evidence bearing on the issue of whether fertility for conscientiousness has been dysgenic in the economically developed nations. First, we have seen that the better educated and higher socioeconomic classes are higher on conscientiousness than the poorly educated and the lower socioeconomic classes. In addition, throughout the economically developed world, the more conscientious classes have lower fertility, so this must inevitably entail dysgenic fertility for conscientiousness. Second, we have seen that psychopaths and criminals, the quintessential criterion groups for low conscientiousness, have larger than average number of siblings, implying that people with low conscientiousness have above-average fertility. Third, we have seen evidence from Britain that criminals have substantially greater fertility than noncriminals, leading to an increase in the number of genes in the population responsible for criminal behavior and hence to an increase in crime. More work is certainly needed on the issue of dysgenic fertility for conscientiousness based on representative samples of the population and on the fertility of criminals compared with that of noncriminals. Nevertheless, taken together the three lines of evidence reviewed in this chapter all point toward the presence of dysgenic fertility for conscientiousness.

The eugenicists feared that individuals with what they called poor character, among whom the criminal class is pre-eminent, were having large numbers of children, that this personality trait is significantly under genetic control, that the genetic quality of the population is deteriorating in this regard, and that the result would be an increase in the rate of crime. Once again, the research evidence has shown they were right.

Chapter 15

Dysgenic Fertility in Economically Developing Nations

1. Latin America. 2. The Caribbean. 3. Asia and the Pacific. 4. Africa. 5. Conclusions.

Hitherto we have been concerned with dysgenic fertility in the economically developed nations. In this chapter we examine the position in the economically developing nations. There have been no studies in these on the relation between intelligence and fertility, but there are quite extensive data for fertility in relation to women's educational level and the socioeconomic status of their husbands. Much of this work has been collated by Ashurst, Balkaran and Casterline (1984) from surveys carried out in the late 1970s, and their results form the principal basis of this chapter, together with the more detailed work in Latin America analyzed by Martin and Juárez (1995). We look first at Latin America, proceed next to the Caribbean, Asia and finally to Africa.

1. LATIN AMERICA

Fertility in relation to women's education and husbands' occupational status for nine Latin American countries for the late 1970s is shown in Table 15.1. The figures are for 20 to 49-year-olds and are age standardized to give estimates of completed fertility. The dysgenic fertility ratios have been calculated for fertility in relation to educational level using the same method as was used for the economically developed nations, namely, by dividing the fertility of the group with the poorest education by that of the highest. It will be noted that all the dysgenic fertility ratios are greater than one, indicating the universal presence of dysgenic fertility. In general, fertility declines in a linear fashion with increasing educational level and socioeconomic status. All of these countries are at the beginning of the demographic transition. The least educated women with

Table 15.1
Women's Fertility in Latin America by Years of Education and Husband's Occupation in the 1970s

	YEARS OF EDUCATION				DYSGENIC	HUSBAND'S OCCUPATION			
	0	1-3	4-6	7+	RATIO	Agric	Skld, unskl	Sales, serv	Prof, cler
Colombia	6.84	6.61	4.68	4.06	1.68	7.44	4.78	4.67	3.96
Ecuador	7.60	7.57	5.94	3.79	2.01	7.60	6.09	5.16	3.95
Paraguay	8.01	6.90	5.43	4.35	1.84	7.30	5.00	4.56	4.41
Peru	7.88	7.26	5.92	5.35	1.47	7.97	6.82	5.94	5.21
Venezuela	7.10	6.30	4.88	3.93	1.81	5.15	5.30	8.02	4.27
Costa Rica	5.84	4.83	3.97	3.98	1.47	5.10	4.11	4.44	3.66
Mexico	7.89	7.63	6.56	4.87	1.62	8.02	7.08	6.41	5.20
Panama	6.56	6.29	4.84	3.91	1.68	6.13	4.39	4.53	4.05
Guyana	6.11	6.01	4.82	5.15	1.19	5.78	5.16	5.04	4.09

Source: Ashurst, Balkaran and Casterline (1984).

Table 15.2
Fertility of Women in Eight Latin American Countries in the Late 1980s in Relation to Years of Education

	Years of Education					Dysgenic
Country	0	1-3	4-6	7-9	10+	Ratio
Bolivia	6.2	6.4	5.3	4.2	2.8	2.21
Brazil	6.7	5.2	3.4	2.8	2.2	3.05
Colombia	5.6	4.5	3.6	2.5	1.8	3.11
Ecuador	6.4	6.3	4.7	3.5	2.6	2.46
El Salvador	6.0	5.2	3.9	3.5	2.5	2.40
Guatemala	6.9	5.6	4.2	2.8	2.7	2.56
Mexico	6.4	6.3	4.0	2.7	2.4	2.67
Peru	7.4	6.1	4.6	3.7	2.5	2.96

Source: Martin and Juárez (1995).

their six to eight children are displaying natural fertility, virtually entirely unchecked by any attempt at family limitation. Women with some education tend to be practicing birth control which is bringing their fertility down—in the case of the most educated, into the 3–5 child range. These figures are confirmed by the second set of data showing women's fertility in relation to their husbands' occupation. Here the wives of agricultural workers are displaying natural fertility with six or more children, except for a suggestion of some control over fertility in Costa Rica, Venezuela and Guyana. The wives of men in professional and clerical occupations are showing some use of family limitation with family sizes in the 3–6 range. The very high figure for fertility of the wives of sales and service workers in Venezuela is probably a sampling or a printing error.

More recent data for women's fertility in relation to educational level have been collected by Martin and Juárez (1995) for eight Latin American countries. These results are for the late 1980s, about a decade later than those shown in Table 15.1, and they provide useful information on women's educational level in relation not only to fertility but also to their desired family size and use of contraception. The data are for all women aged 15 to 49 and are calculated as total fertility rates. Fertility in relation to five educational levels is shown in Table 15.2. There are three points of interest in the table. First, natural fertility of six or more children is still generally present among women with no education, with the exception of a slight fall to below six in Colombia. Second, fertility declines steadily and in linear fashion with increasing educational level

Table 15.3
Women's Desired Number of Children in Relation to Years of Education in Eight Latin American Countries in the Late 1980s

	Years of Education				
Country	0	1-3	4-6	7-9	10+
Bolivia	2.6	2.6	2.9	2.8	3.0
Brazil	3.0	3.1	2.9	2.9	2.9
Colombia	3.2	3.1	3.0	2.9	2.9
Ecuador	3.4	3.4	3.3	3.0	3.1
El Salvador	4.3	4.0	3.5	3.3	3.3
Guatemala	4.6	4.0	3.5	3.7	3.4
Mexico	4.0	3.4	3.2	3.0	2.9
Peru	3.0	3.1	2.8	2.8	2.8

Source: Martin and Juárez (1995).

until it is well-controlled among the most educated women who have between 1.8 and 2.8 children. Dysgenic fertility in Latin America in the late 1980s was pronounced.

Martin and Juárez also provide data for desired family size and these are shown in Table 15.3. It will be noted that the desired family-size differences in relation to educational level are very small. In Bolivia the best educated women actually have a greater desired family size than the least educated (3.0 versus 2.6); in Brazil, Colombia, Ecuador and Peru the differences are small or negligible; in El Salvador, Guatemala and Mexico the differences are a little larger but nowhere near as great as the fertility differences.

How does it come about that there are only small differences in desired family size in relation to educational level in these countries, but the actual family-size differences in relation to educational level are so much greater? The answer to this question is that educated women use contraception more and control their family size more effectively. This is shown in Table 15.4 which gives the use of contraception for married women in relation to educational level. Notice that the use of contraception increases in generally linear fashion with increasing levels of education and is much greater among the most educated as compared with the least educated. It is often asserted that the poor in economically developing countries have large numbers of children to work on the land, and to keep them in their old age. The data set out in Tables 15.3 and 15.4 suggest this is incorrect. The conclusion to be drawn from these tables is that dysgenic

Table 15.4
Women's Use of Contraception (Percentages) in Relation to Years of Education in Eight Latin American Countries in the Late 1980s

	Years of Education				
Country	0	1-3	4-6	7-9	10+
Bolivia	12	23	31	43	53
Brazil	47	59	71	76	73
Colombia	53	61	65	73	73
Ecuador	18	37	43	50	61
El Salvador	37	42	55	51	64
Guatemala	10	24	42	60	60
Mexico	25	44	58	70	69
Peru	19	33	46	60	67

Source: Martin and Juárez (1995).

fertility is present in Latin America largely because of educational differences in the efficient use of contraception and not because there is a strong inverse relationship between women's educational level and their desired family size. Dysgenic fertility is present in Latin America mainly for the same reason that it is present in the economically developed nations of North America and Europe.

2. THE CARIBBEAN

Women's fertility findings in relation to their education and husbands' occupational status for four Caribbean countries are shown in Table 15.5. In the Dominican Republic and Haiti women with zero and with 1.3 years of education, and those whose husbands are agricultural workers show natural fertility with six to eight children, while the most educated and the wives of professional and clerical men have four to five. In Jamaica and Trinidad and Tobago, dysgenic fertility is also present but is less pronounced.

3. ASIA AND THE PACIFIC

Similar data for women's fertility in relation to their educational level and their husbands' socioeconomic status are shown for fifteen countries of Asia and the Pacific in Table 15.6. The results for China come from a survey carried

Table 15.5
Women's Fertility in the Caribbean by Years of Education and Husband's Occupation in the 1970s

	Years of Education				Dysgenic	Husband's Occupation			
	0	1-3	4-6	7+	Ratio	Agric	Skld, Unskld	Sales, Serv	Prof, Cler
Dominican Republic	6.72	7.00	5.61	4.60	1.46	7.69	5.40	5.23	4.04
Haiti	7.04	6.08	5.35	4.57	1.54	7.31	5.90	5.90	4.64
Jamaica	5.85	4.90	5.25	4.78	1.22	5.98	5.06	3.93	3.71
Trinidad and Tobago	4.40	3.40	3.79	3.50	1.26	4.41	3.78	3.32	2.81

Source: Ashurst, Balkaran and Casterline (1984).

Table 15.6
Women's Fertility in Asia by Years of Education and Husband's Occupation in the 1970s

	YEARS OF EDUCATION				DYSGENIC	HUSBAND'S OCCUPATION			
	0	1-3	4-6	7+	RATIO	Agric	Skld, unskl	Sales, serv	Prof, cler
Jordan	9.04	7.97	6.96	6.26	1.44	9.22	8.41	8.44	6.93
Syria	9.08	7.29	6.14	6.04	1.50	9.38	8.31	7.21	7.90
Turkey	5.41	4.07	3.30	3.22	1.68	5.41	4.24	3.98	3.65
Yemen AR	8.36	----	6.13	----	----	7.85	8.87	8.79	7.30
Bangladesh	5.43	5.57	5.93	5.42	1.01	5.55	5.30	5.71	5.35
Nepal	6.04	5.83	6.08	3.05	1.98	6.21	5.93	5.66	4.37
Pakistan	6.39	5.80	6.52	4.89	1.31	6.29	6.63	6.46	6.02
Sri Lanka	5.28	5.11	5.29	5.30	0.99	5.42	5.22	5.41	4.96
Fiji	4.35	4.30	4.94	4.68	0.93	5.16	4.52	4.30	4.22
Indonesia	4.60	5.30	5.20	5.33	0.86	4.51	5.26	5.27	5.07
Korea, Rep	6.22	5.58	5.45	4.97	1.25	6.39	5.23	4.81	5.10
Malaysia	5.89	5.74	5.54	5.22	1.13	5.91	6.02	5.60	5.00
Philippines	6.75	7.54	7.10	5.99	1.13	7.38	6.65	5.92	5.69
Thailand	5.59	5.75	5.62	4.47	1.25	6.04	5.23	4.73	4.04
China	5.40	5.00	4.30	3.30	1.63	----	----	----	----

Source: Ashurst, Balkaran and Casterline (1984).

out in 1987 and reported by Zhihui (1990) and are for women born in approximately the first quarter of the twentieth century. The countries are divided into three groups consisting of four in the Near East, four in the Indian subcontinent, and seven in East Asia and the Pacific.

Among the four Near East countries, Jordan, Syria and the Yemen are evidently at the very beginning of the demographic transition in which natural fertility of eight to nine children is present among the least educated women, while the most educated are having six children. This suggests a slight element of control of fertility among the most educated, as occurred in the United States and Europe at the start of the demographic transition in the early and middle decades of the nineteenth century. Wives' fertility in relation to husbands' occupation tells the same story with somewhat higher fertility among the wives of agricultural and unskilled workers as compared with that of professional and clerical classes.

Turkey is an exception among this group of countries. Fertility is below six in all educational and occupational classes, but is rather considerably more controlled among the better educated and the wives of professional and clerical men, giving rise to a high dysgenic fertility ratio.

Turning to the four countries of the Indian subcontinent, Nepal and Pakistan show the usual pattern of high overall fertility and fairly strong dysgenic fertility ratios of 1.98 and 1.31, respectively. Bangladesh and Sri Lanka show an unusual pattern in which fertility is more or less uniform at about five-plus children in all educational and occupation groups; hence, there is no dysgenic fertility. The figures suggest that there is some control over fertility which is spread fairly uniformly over all educational and occupational groups.

The seven countries of East Asia and the Pacific present some further contrasts. In Fiji and Indonesia fertility appears to be under some degree of control, but there are not the usual differentials in relation to educational level or socioeconomic status. Indonesia is particularly anomalous, the average fertility of around five suggesting some control over fertility but a complete absence of the usual dysgenic pattern. In fact, fertility is positively associated with both the educational level of women and the occupational status of their husbands, recalling the pattern of preindustrial societies that we examined for Europe in Chapter 2.

The remaining five countries display the fertility pattern typical of the early stage of the demographic transition, with close to natural fertility among the least educated women and the wives of agricultural workers and the skilled and unskilled, and some degree of fertility control among women with better education and the wives of professional and clerical men, which results in the familiar dysgenic fertility ratios.

4. AFRICA

The continent in which least progress has been made toward the demographic transition is Africa. The fertility data for thirteen countries are shown in Table

15.7. Notice that fertility is generally very high, indicating either natural fertility or close to it in all educational and occupational categories, among whom the number of children are in excess of 5.0, except for the most educated women in Egypt and Tunisia.

In five of the countries—Benin, Ghana, Ivory Coast, Senegal and Sudan—the dysgenic fertility ratios are in the range of 1.11 to 1.21. These figures indicate the first beginnings of the demographic transition consisting of a slight reduction of fertility among the most educated. In another five countries—Cameroon, Kenya, Lesotho, Nigeria and Morocco—there is virtually no dysgenic fertility. It is only in the two northern countries of Egypt and Tunisia that the demographic transition shows significant signs of having begun among the most educated class, and consequently where dysgenic fertility is appreciable. More recent data for Morocco collected in a social survey carried out in 1992 show the onset of considerable dysgenic fertility in relation to women's educational level. The average number of children were 4.9 for those with no education, 2.4 for those with primary education and 2.0 for those with secondary education (Azelmat, Ayad and Housni, 1992).

It will be noted that in seven of the eight countries of sub-Saharan Africa, fertility increases slightly from women with no education to those with one to three years, and then falls slightly among those with more education. Probably the explanation for this is that women with no education are largely in rural areas where there are some traditional fertility limitation conventions, such as abstinence from sexual intercourse by women who are breast-feeding. Those with one to three years education are more commonly in urban areas where these conventions have broken down, with the result that their fertility is higher.

The high fertility generally present in all educational and social classes in sub-Saharan Africa suggests negligible practice of family limitation. This inference is confirmed by studies of the use of contraception which show very low usage. Sonko (1994) in a review of this question concludes that contraception usage is about 5 percent in sub-Saharan Africa, except for Botswana, Kenya and Zimbabwe, where it is a little higher. In countries where the average fertility of women is below seven, this is not because of the use of contraception but because of the widespread prevalence of infertility, much of it due to sexually transmitted diseases. Sonko gives an account of the attempt by the government of Ghana to introduce a population policy in 1969, aiming at the reduction of fertility. Many family planning clinics were set up throughout the country but by 1987 the fertility rate was the same as it had been eighteen years previously. For some reason, noted by demographers but not understood by them, family limitation has not been adopted in sub-Saharan Africa, as it was in Europe and North America between the years 1870–1920, and has been adopted more recently in North Africa, in Asia and in Latin America. If the demographic transition eventually occurs in sub-Saharan Africa, it is likely to begin first among the more educated and higher social classes, some early signs of which are

Table 15.7
Women's Fertility in Africa by Years of Education and Husband's Occupation in the 1970s

	YEARS OF EDUCATION				DYSGENIC	HUSBAND'S OCCUPATION			
	0	1-3	4-6	7+	RATIO	Agric	Skld, unskl	Sales, serv	Prof, cler
Benin	6.80	7.95	6.20	5.64	1.20	6.78	7.05	6.04	6.79
Cameroon	5.59	6.57	6.50	5.26	1.06	5.76	5.77	6.66	5.82
Ghana	6.48	6.99	7.02	5.66	1.14	6.56	6.74	6.36	5.48
Ivory Coast	6.75	6.91	5.82	5.65	1.19	6.91	6.53	6.48	5.64
Kenya	7.43	8.53	7.90	7.83	0.95	7.79	7.73	8.13	7.79
Lesotho	5.76	5.45	6.00	5.96	0.97	6.12	5.91	5.59	6.52
Nigeria	5.69	5.85	7.68	5.40	1.04	5.53	7.19	5.98	6.34
Senegal	6.64	8.96	6.42	5.96	1.14	6.62	6.93	6.37	7.04
Egypt	6.13	5.90	5.63	4.96	1.24	6.51	5.67	5.72	5.07
Mauritania	6.56	----	6.47	----	----	6.41	6.54	6.44	7.33
Morocco	6.70	5.86	5.86	6.17	1.09	7.36	6.46	5.68	5.37
Sudan	6.42	6.23	5.82	5.31	1.21	6.45	6.49	6.44	5.56
Tunisia	7.53	6.42	6.06	4.67	1.61	7.74	6.66	7.05	5.51

Source: Ashurst, Balkaran and Casterline (1984).

already beginning, and consequently will entail the dysgenic fertility and genetic deterioration that has occurred in the economically developed nations.

5. CONCLUSIONS

We have seen that the inverse relationship between educational level and fertility, and between socioeconomic status and fertility, is generally present throughout much of the economically developing world as it enters the early stage of the demographic transition. For this to be dysgenic there would have to be some degree of the same genetic segregation in relation to educational level and social class that we have seen in the economically developed nations. There is no direct evidence that this is the case, but wherever there is social mobility there is inevitably some genetic segregation by intelligence and conscientiousness, the two major determinants of social mobility. All societies have some social mobility, so some genetic segregation by educational level and social class must be present in the economically developing world. It follows that throughout most of this world genetic deterioration must be taking place. There are, however, two exceptions to the generally prevailing dysgenic fertility. First, there is little sign of dysgenic fertility in sub-Saharan Africa. The reason for this is that family limitation has hardly begun to be practiced in any strata of society, so educational and socioeconomic fertility differentials are either minimal or nonexistent. The second exception consists of four countries in the Asia and Pacific group, namely, Bangladesh, Sri Lanka, Fiji and Indonesia, where the moderate fertility trends suggest some degree of control over family size, but the usual differentials in relation to educational and socioeconomic status are not present. The apparent failure of dysgenic fertility to appear in these countries is an anomaly which deserves further research. These exceptions apart, it is clear from the evidence reviewed in this chapter that dysgenic fertility, and the genetic deterioration it implies, has not been confined to the economically developed world. On the contrary, it is virtually a worldwide phenomenon of modern populations in the twentieth century.

Chapter 16

Counterarguments, Rejoinders and Conclusions

We saw in the first chapter of this book how the eugenics movement was founded by Francis Galton and won many adherents in the first half of the twentieth century and that, in the second half of the century, eugenics came under increasing attack. This book has been devoted to assembling the evidence for the eugenicists' view that the genetic quality of the populations of the Western nations is deteriorating. Why, then, did the eugenicists' analysis come to be repudiated and what were the arguments against it? This is the question we take up in this concluding chapter.

1. QUESTIONING GENETIC EFFECTS

The case for eugenics depends crucially on the axioms that intelligence, conscientiousness, educational achievement and social class are significantly determined by heredity. If they were not, the inverse associations between these characteristics and fertility would not pose a genetic problem. A number of the critics of eugenics have mounted attacks on this component of the eugenic argument. For instance, Daniel Kevles writes of ''the methodological ricketiness in hereditarian theories of intelligence'' (1985, p. 395) and Carl Degler that ''social scientists doubted the validity of eugenics because of its misplaced faith in the assumption that mental traits could be inherited'' (1991, p. 146). In the case of intelligence, the leading critic of the view that it has no significant heritability has been Leon Kamin (1974), but his arguments have been universally rejected by experts in this field, such as Mackintosh (1975), Eysenck (1979), Vernon (1979), Jensen (1980), Plomin, De Fries and McClean (1990), Brody (1992) and Bouchard (1993). Even Steve Jones (1993) admits that intelligence has a significant heritability. The evidence that intelligence has high heritability is accepted by all scholars who have given their attention to this question, with the single exception of Kamin, and is set out in standard textbooks, such as those of Eysenck (1979) and Brody (1992).

The critics of eugenics have also disputed the evidence for the heritability of conscientiousness and crime. Jones writes that "the link between genes and crime is such a distant one as almost to lack meaning"—almost, but perhaps not quite—and of the "the eugenical cranks who believed that antisocial behavior was coded in the genes" (1993, p. 238). But Jones does not mention, let alone attempt to refute, the numerous twin and adoption studies showing that crime has a high heritability, reviewed in Chapter 13. Other critics of eugenics ignore the whole issue of the genetic basis of crime. For instance, Steven Rose, Leon Kamin and Richard Lewontin (1984) in their book *Not in Our Genes*, a thorough attack on theories of the genetic determination of human behavior, make no mention of this extensive evidence, although they devote a couple of pages to criticizing the hereditarian views of Emile Zola, the French nineteenth century novelist. Zola was an easy Aunt Sally, but they ducked any confrontation with the hard evidence for a high heritability of crime based on twin and adoption studies.

The heritability of educational attainment has also been attacked. When in 1983 Lee Kwan Yew, the prime minister of Singapore, expressed his concern about the small number of children of women graduates and asserted that this was having a dysgenic effect because educational level is associated with IQ and has a genetic basis, he was ridiculed by Stephen J. Gould: "Nothing strikes me as more silly or self-serving than the attempt to infer people's intrinsic, genetically based intelligence from the number of years they attended school" (1984, p. 22). Gould was apparently unaware of the vast amount of evidence that intelligence and educational level are highly correlated; that intelligence has a high heritability which must be present also in educational attainment; and that numerous studies have shown that educational attainment itself has a substantial heritability, which have been reviewed in Chapter 9. Gould not only asserts that educational attainment has no heritability but also believes that no human behavior is genetically determined: "What is the direct evidence for genetic control of specific human behavior?" he asks and concludes, "At the moment, the answer is none whatever" (1976, p. 345). The only conclusion that can be drawn from this statement is that Gould has never bothered to read the research literature on these issues.

The eugenicists believed that intelligence, mental retardation, conscientiousness, crime, alcoholism and educational attainment all have substantial heritabilities. By the 1990s there was a massive accumulation of evidence in support of this view. Some of the critics, like Kamin, who disputed this did so by using the nit-picking methods that Bouchard (1993) has rightly called "pseudo-analysis." Most of the others, like Jones, Lewontin, Rose and Gould, simply ignored it. Their repeated assertions that these human traits have no heritability rest on nothing more substantial than shoddy scholarship.

2. MUST INTELLIGENCE BE DETERIORATING?

Evolution takes place through some individuals having greater fertility and lower mortality than others, with the result that the traits of these individuals increase in the population. Two attempts have been made to show that this fundamental law of biology does not hold for the trait of intelligence in human populations. The first of these was by Penrose (1948, 1950), who proposed a genetic model in which intelligence is determined by a single gene with two alleles, *A* for high intelligence and *a* for low. These produce three genetic varieties. The *AA*s have high intelligence because they have the two high-IQ alleles. Penrose's model posits that this confers an IQ of 108 and that 86 percent of the population belong to this group. The second group is *Aa*, with one allele for high intelligence and one for low, consists of 8 percent of the population and has a mean IQ of 66. The third group, *aa*, has both the low-intelligence alleles, has a mean IQ of 24, and consists of 6 percent of the population. The mean IQ of the total population works out at 99.6.

Penrose posited that the *AA* group has an average fertility of 1.9 children, the *Aa* of 4.0 children, and the *aa* are infertile. He argues that the generation of children will have the same IQ as that of the parents because *Aa* × *Aa* matings will produce one *AA*, two *Aa* and one *aa* children, thereby contributing 25 percent of *AA*s to the child generation and keeping the intelligence level of the total population stable. In general terms, the argument is that the relatively low fertility of the high-IQ *AA*s is counterbalanced by the infertility of the low-IQ *aa*s.

There are two problems with the argument. First, the posited distribution of intelligence and infertility bears little resemblance to that actually present. The model depends on 6 percent of the population with a mean IQ of 24 being childless. In the real world the severely mentally retarded in the IQ range 0–50 and with an average IQ of 24 are not always childless, as shown in Bajema's sample in Table 6.3; and even if they were, they comprise only about 0.5 percent of the population and are too few to have a significant effect on the overall inverse relationship between intelligence and fertility.

A second problem with Penrose's model is that because 86 percent of the population are *AA* and therefore genetically identical, there would be very low genetic variance for intelligence. The effect of this would be that heritability studies would show much lower heritabilities than the .80 or so that are actually obtained. For instance, in studies of identical twins reared apart, 86 percent of the pairs would be *AA*s, so differences between their IQs would be entirely environmentally determined. The remaining 14 percent would add some genetic variability, but nowhere near sufficient to produce the high twin-twin correlations that are actually obtained. Similarly, the correlations for identical and fraternal twins would be much more similar than those actually found. Nonidentical twins would be genetically identical in 86 percent of cases. So their IQs would be closely similar and the correlation between them much higher than the .5 typically found. Penrose's model is seriously inconsistent with the research evidence on the size of an infertile low-IQ subgroup and on the heritability of intelligence.

A more recent attempt to show that the inverse relationship between intelligence and fertility does not necessarily entail a decline of genotypic intelligence has been made by Preston and Campbell (1993). They posit a model in which there are a number of IQ classes and people only mate within their IQ class (endogamous mating). They present a mathematical model which purports to show that differential fertility for intelligence can produce either a rise or decline in the intelligence of subsequent generations, and in both cases eventually reaches equilibrium.

There are several problems with the model. Perhaps the most important is that it attempts to prove mathematically that Darwin's theory of evolution does not work. Evolutionary theory maintains that if those with a certain trait have more offspring who survive to adulthood than those without the trait, the trait increases in the population in each succeeding generation and eventually those lacking in the trait are eliminated. Preston and Campbell understand this but say that "the intuition that differential fertility will change population characteristics appears to be based on the assumption that a population consists of closed subpopulations, each of which reproduces at its own rate . . . but a strict analogy to isolated populations is incorrect in the case of IQ and many other traits, since changes in scores from parents to child are common" (p. 998). This is a misunderstanding of Darwinian evolution, which does not require closed subpopulations to work, although it does sometimes operate at this level. Darwinian evolution frequently works through differential fertility in open populations, such as exist for intelligence. What happens in natural conditions is that from time to time an individual is born with a new mutant gene which confers a reproductive or survival advantage. This individual has greater than average surviving children, so the gene gradually spreads in the population. This occurs in open populations and does not require the closed subpopulation envisaged by Preston and Campbell in which one subpopulation has the gene and the other does not. If the Preston and Campbell model were correct, evolution would not work. There are also problems with the assumptions on which the model is based. The assumption that people only mate endogamously, within their own IQ class, is invalid because in the real world many people mate outside their IQ class. This is shown by the relatively low correlation of about .33 between the intelligence of married couples (Brody, 1992). Another assumption in the model which is questionable is that in the posited endogamous mating system, descendants of all IQ classes could appear in any other class, that is, descendants of parents both of whom are in the class of IQ 126+ could appear in the lowest IQ class of below 75. This is improbable because it would require much greater regression effects than are observed. For instance, in the data set which Preston and Campbell use for their samples, parents with IQs of 126+ have children with a mean IQ of 123.5 and variance of 13.0. With figures like these, the probability of these parents having descendants with IQs below 75 is negligible, except for occasional mutations and adverse recessive gene combinations which would be too rare to have significant effects. Although there is a small regression effect in the first generation, this would not be expected to continue in subsequent generations. Other questionable assumptions in the model are discussed by Coleman (1993) who concludes that the model is invalid and that the mainstream view that dysgenic fertility leads to a deterioration of genotypic intelligence is correct. Selective breeding has been demonstrated by animal and plant breeders for centuries to produce improved strains. If the Preston-Campbell model were correct, this would not be possible. It cannot be argued that intelligence is an exception to this principle.

3. GENETIC DIFFERENCES BETWEEN THE SOCIAL CLASSES

None of the axioms of eugenics has been so strongly disputed as the existence of genetic differences between the social classes. The eugenicists believed that the professional class is a genetic elite, although they recognized that it is an open elite, which in each generation recruits the gifted from other classes and expels the less gifted from its own ranks.

This proposition was attacked by the historians of eugenics. Haller (1963) asserted that there was ''almost no evidence'' for it, perhaps hedging his bets by the implication that some evidence did exist. In the next decade Ludmerer (1972) wrote that ''the English Eugenics Education Society largely consisted of a collection of social snobs very conscious of 'being well-born' who expressed silly views on hereditary'' (p. 69). According to these critics, the belief that the social classes differ genetically was mere social snobbery. The argument was reasserted by Kevles, who writes that ''there was no evidence that the higher birth rate of lower income groups was polluting the gene pool'' (1985, p. 285), which it would have done if the lower social classes were genetically inferior.

It was not only historians who asserted that there are no genetic differences between the social classes. A number of biologists and geneticists were equally confident that this was the case. The American geneticist Theodosius Dobzhansky discussed this issue in 1959 in his Silliman lecture delivered at Yale. He asserted that there were no genetic social-class differences in ability or talent in Western societies. Such differences could not have arisen, he stated, because there has not been equality of opportunity over a sufficient period of time to allow the social mobility necessary to bring them into existence. ''The occupational and class differentiation in the existing societies is not established on a genetic basis, precisely because none of these societies now provides, or has provided, anything approaching an equality of opportunity to its members'' (1962, p. 248). Curiously enough, a few pages earlier Dobzhansky stated that there has been social mobility in Western nations and that ''even the most rigid class society allows some individuals of humble birth to climb and others of privileged birth to slide down the social ladder'' and three pages later we find him writing of ''the relatively open-class society of America'' (pp. 242, 251). Dobzhansky failed to think through the fact that even a small element of social mobility over a number of generations must inevitably lead to some degree of segregation of the genes for high intelligence and conscientiousness in the professional and middle classes. Furthermore, the degree of social mobility has been substantial over many centuries in Western nations and also in China, as we saw in Chapter 2. Dobzhansky was clever enough to see that to deny the existence of genetically based social-class differences in intelligence it is necessary to deny that there has been social mobility, although he himself realized that there has been social mobility in Western societies for many centuries.

Other biologists who denied the existence of genetic differences between the classes were Milo Keynes in Britain and Richard Lewontin in the United States. Keynes asserted in a lecture on Francis Galton that Galton got it all wrong about eugenics because ''there is no class of individuals who are an elite'' (1993, p. 23), while Lewontin confidently asserted that ''there is not an iota of evidence that social classes differ in any way in their genes'' (1993, p. 37). Doubts about the existence of genetically based social-class differences in intelligence were also expressed by the anthropologist Nicholas Mascie-Taylor, who reviewed some evidence showing that social-class differences in intelligence were present and concluded that these ''*might* reflect genetic differences between the social classes, but equally the differences could reflect untested environmental variables,'' and therefore that the issue of ''any underlying genetic differences remains so far largely unresolved'' (1990, pp. 134, 140). None of these critics attempts to refute, or even mention, the seven studies reviewed in Chapter 11, showing that the intelligence of adopted children is positively associated with the social class of their fathers, demonstrating that intelligence differences between the social classes must have a genetic basis.

The more thoughtful and better-informed biologists and social scientists, even those in the liberal camp, concede that this is the case. Sir Walter Bodmer and Luigi Cavalli-Sforza (1976) write that ''probably both genetic and cultural differences matter'' in the attainment of socioeconomic status; and Sandra Scarr has sorrowfully concluded that the social classes differ genetically, writing that ''I am reluctantly persuaded that within the white population in the United States, there are average genetic differences among the offspring in working, middle and upper-middle-class families and that these genetic differences explain half or more than half their average intellectual differences'' (Scarr, 1984, p. 527).

The eugenicists believed that the lower fertility of the professional class was a cause for concern because the professional class is a genetic elite and, more generally, that there is a genetically based social-class gradient for intelligence and character. The evidence for this belief has been examined in Chapters 11 and 13, where it was shown that it has been confirmed by adoption studies carried out in the United States, Britain, Denmark and France. All those who have examined the evidence accept this, although some have done so reluctantly. The only conclusion that can be drawn regarding those who continue to dispute this is that they have never bothered to look at the evidence.

4. NO BAD GENES

The eugenicists thought that the genes responsible for genetic diseases, low intelligence and poor conscientiousness are undesirable both for the individual carrying them and for society. In the last quarter of the twentieth century this belief came to be challenged by critics of eugenics, such as G.W. Lasker (1991), a biologist at Wayne State University in Detroit, who writes that "the eugenics movement, based on the idea that some genes are good and others bad, has been discredited" (p. 2).

There are several variants of the argument that there are no bad genes. The first affirms that some genes are advantageous in particular environments but disadvantageous in others, so there is no such thing as a bad gene in any absolute sense. This argument was advanced by Dobzhansky (1962, p. 288) in his assertions that "A gene harmful in one environment may be neutral or useful in another" and "What is good in the Arctic is not necessarily good in the tropics; what is good in a democracy is not necessarily good under a dictatorship."

This argument is valid for certain physiological and morphological characteristics. For instance, it is better for an animal to have white fur in the Arctic because it provides camouflage in a snowy environment, but not so good in the tropics where it would make an animal conspicuous. Conversely, brown fur is better in the tropics, but not in the Arctic. So we can accept that there is no such thing as a good or bad gene for the fur color for bears or foxes. It all depends on where they live.

For the extension of this argument to humans, the classical example is the sickle cell anemia gene which provides immunity against malaria when carried as a recessive in so-called heterozygotes, and is to this extent a good gene. However, it produces the severely crippling disease of sickle cell anemia in homozygotes, individuals who inherit the double recessive from both parents. It can therefore be argued that the sickle cell gene is neither good nor bad. It is good when carried as a single recessive in places where malaria is prevalent, but bad when inherited as a double recessive.

The argument only applies, however, in environments where there is malaria. In environments where there is no malaria, such as in New York or London, the sickle cell gene confers no advantage when carried as a recessive and produces a very unpleasant disorder when inherited as a double recessive. In these environments the sickle cell gene is wholly a bad gene. Furthermore, the sickle cell anemia gene is a very unusual example of an apparently harmful gene which is beneficial in malarial environments when inherited as a single recessive. It is generally believed that many of the recessive genes for various diseases arose and spread because at some time in the past they provided immunities against diseases which have long since disappeared. These genes have persisted because single recessive genes generally do no harm, so it takes a very long time for them to be eliminated from the population. These genes are, nevertheless, wholly bad genes because of the serious disabilities they cause in children who inherit them from both parents.

The extension of the argument that there is no such thing as a bad gene to human intelligence and conscientiousness is equally lacking in credibility. Dobzhansky seems to envisage a range of human environments, for some of which high intelligence and conscientiousness are advantages, but not others. This seems to be his meaning when he asserts that some genes are advantageous for people living in democracies, while other genes are better in dictatorships—but he does not specify what these are. They cannot be the genes for intelligence because the genes for low intelligence contribute to the carrier's poor educational and occupational attainment; to unemployability and to crime; and they seriously impair the capacity for making a positive contribution to society. These are poor genes for all societies—from primitive hunter-gatherers to modern democracies and dictatorships. The only reasons they persist are that they have not yet been bred out of the population; they continue to appear through mutation; and in the course of the last century and a half, they have been spreading in the economically developed nations as a result of the high fertility of the less intelligent. Genes for low intelligence are bad genes and it would be desirable to eliminate them.

The issue of bad genes with regard to conscientiousness is a little more complex. In most environments it is obviously better to be conscientious than unconscientious, because conscientiousness contributes to educational and occupational attainment, while low conscientiousness is a major contribution to crime, imprisonment and social ostracism. There may perhaps be a few environments in which low conscientiousness is an advantage. The Marine Corps and the French Foreign Legion are possible examples. It is the job of those who serve in these forces to kill people when the occasion demands, and a too highly developed conscience would probably impair the discharge of this responsibility. Nevertheless, these are unusual environments, and even in these, very low conscientiousness in the form of psychopathic personality would be disadvantageous because of the need for military discipline and group cooperation. It may be that Dobzhansky had conscientiousness in mind when he asserted that some genes are advantageous in democracies and others in dictatorships, and he might have argued that low conscientiousness would serve the individual better in a dictatorship. Alternatively, it may well be that individuals with low conscientiousness are much more likely to be shot in dictatorships. Dobzhansky's argument needs spelling out if it is to be rendered convincing, and it is a pity he did not take the trouble to do this.

5. CREATIVITY AND GENETIC DISEASE

A further variant of the argument that there are no bad genes is that genetic diseases make a positive contribution to creative achievement. One of the protagonists of this argument is George Steiner who has asserted that "What in many cases is a hideous disease, a handicap, can also be profoundly creative. Without the kind of meningital deafness which comes of inherited syphilis and alcoholism you and I would be sitting here without Ludwig van Beethoven" (Harris, 1992, p. 84). In further support of this argument, Steiner cited the case of the disabled painter Henri Toulouse-Lautrec, whose disability he attributed to the genetic disease muscular dystrophy. The genius of this painter "sprang out of very profound physical handicaps." Steiner's medical knowledge is rather sketchy; it is unlikely that Toulouse-Lautrec had muscular dystrophy because nearly all of those who suffer from this disease are unable to walk by early adolescence and die before their mid-twenties (Gilbert, 1993). The standard biographies of Toulouse-Lautrec state that his disabilities were caused by a fall at the age of fourteen in which he broke his leg (Thorne and Collocot, 1984). This point aside, there is no evidence that genetic diseases contribute in any way to creative genius or outstanding achievement. If these speculations contained any truth, it would be possible to demonstrate that these genetic diseases were frequently associated with creative genius, but this has never been done. They are simply chance coincidences and there is no persuasive evidence to suggest that they make any contribution to creative achievement.

The research evidence on this issue indicates that, far from having poor health, creative geniuses and outstanding achievers have better than average health. Post (1994) has made a study of the life histories of 291 of the most illustrious men of the nineteenth and early twentieth centuries and concludes that they were much healthier than their contemporaries. Their average life span was 68 years, well above the population mean for this time, and only 8 percent had debilitating illnesses. In addition, the classical study by Terman (1959) of approximately 1,500 high-IQ and high-achieving individuals from California found that they were well above average with regard to physical health.

The critics of eugenics have asserted that there is no such thing as a bad gene. Try telling this to someone who has inherited the genes for cystic fibrosis, Huntington's chorea or one of the other four thousand or so genes for debilitating disease; or to the parents of a child with mental retardation or psychopathic personality. The eugenicists believed there are bad genes that would be best eliminated. They were right.

6. VALUE JUDGMENTS ABOUT HUMAN QUALITY

The eugenicists thought it was better to be healthy than diseased, better to be intelligent than dull and better to be conscientious than psychopathic. These beliefs also came under attack in the second half of the twentieth century. Thus James Neel, a population geneticist at the University of Michigan, writes that "Any attempt at varying the number of children according to parental attributes requires massive value judgements which cannot be supported on social or scientific grounds" (1994, p. 340). David Smith (1994), an educational psychologist at the University of South Carolina, asks, "Is mental retardation always a disease to be prevented, or is it a human condition worthy of being valued?" It is a rhetorical question implying that the mentally retarded are just as valuable to society as the intelligent. Others went further and suggested that it was actually better to be stupid than to be intelligent. In Britain Donald McKay warned against any attempt to raise the intelligence level of the population on the grounds that if this were done criminals would become more intelligent, would practice their profession more efficiently, be more difficult to catch and become an increasing social menace (1963, p. 298). He does not seem to have known that criminals typically have low IQs and are attracted to crime partly because they cannot acquire the vocational skills to earn good incomes, and partly because they cannot figure out the probabilities of being caught and imprisoned and the long-term adverse consequences of a criminal career. If criminals became more intelligent they should be able to earn good incomes by legitimate means and a criminal career should be a less attractive alternative. Furthermore, if the intelligence level of the population increased, the IQs of the police would be higher and they would be better at catching the criminals.

Another variant of the view that stupidity is better than intelligence is that intelligent people do more harm than the unintelligent. For instance, Leon Kass, a biochemist at the University of Chicago, has maintained that it was the intelligent rather than the mentally retarded who started the Vietnam war, and suggested that the "crusaders against genetic deterioration are worried about the wrong genes" (1972, pp. 8–10), implying that the fewer intelligent people there are the better. A similar view was advanced by the British physician Milo Keynes who argued in a lecture, ironically celebrating the achievements of Francis Galton, that "for the tragedies of mankind, that army of ineffectives, the intellectual proletariat, are blameless" and that human progress is impaired, not by the unintelligent, but by "the stupidity of the intelligent" (1993, p. 23).

These assertions that all human qualities are equally valuable cannot be taken seriously. At the extremes, the unintelligent and the psychopathic are unemployable; make no positive contribution to society; and impose social costs on others, who have to maintain them and, frequently, endure their criminal and other forms of antisocial behavior. There are also social costs involved in the treatment of those with genetic diseases, quite apart from the misery of the individuals who suffer from them, and of their families. It is difficult to believe that these people who assert that all human types are equally valuable, or even that the dull are to be preferred over the intelligent, have given any serious thought to these questions. We spend considerable sums on the education of children and on the attempt to rehabilitate criminals precisely because we believe that it is good to have well-developed cognitive skills and law-abiding citizens. These assertions that stupidity is just as good as intelligence, and psychopathic personality just as good as conscientiousness, are just a form of contemporary cant, reiterated by people who have not taken the trouble to think their position through.

7. MAGNITUDE OF THE PROBLEM

Although the eugenicists were undoubtedly correct in their belief that genetic deterioration is taking place in Western nations, it is worthwhile considering whether this is a serious or only a trivial problem. With regard to the deterioration of health, the magnitude of the problem is difficult to quantify. The birth incidence in Western populations of the single-gene disorders, such as phenylketonuria, cystic fibrosis, muscular dystrophy, hemophilia and so on, is about four per thousand, and of the multifactorial disorders in which genetic processes are involved, such as spina bifida, is about 46 per thousand (Baird, Anderson, Newcombe and Lowry, 1988). Many of those born with these disorders die in infancy and childhood or, if they survive into adulthood, do not have children, so they are not a dysgenic problem. But medical progress has brought treatments for a number of these conditions, enabling those affected to survive and have children, with the result that the incidence of the disorders is doubling or tripling in each generation (Modell and Kuliev, 1989). Medical advance in the treatment of these disorders is rapid and will inevitably increase their incidence in the future.

To some degree this increase is being mitigated by the prenatal diagnosis and abortion of fetuses with these disorders—a form of eugenic intervention which has generally escaped the censure of the critics of eugenics, most of whom have been able to see that it is better to be born healthy than diseased. There are some 4,000 single-gene diseases, as well as a number of multifactorial genetic disorders. The overall impact on their future incidence cannot be estimated with any precision because of advances in the treatment of some of them and by the prenatal diagnosis and abortion of affected fetuses in the other cases. Undoubtedly some of these disorders, such as cystic fibrosis and phenylketonuria, are increasing significantly and this is a problem that needs to be taken seriously.

With regard to intelligence, the emergence of an inverse relationship between socioeconomic status and fertility in the cohorts born between 1815 and 1825 in England indicates the onset of genotypic deterioration—which was probably present in other Western nations and has been documented for many of them from the middle decades of the nineteenth century onward. This means that dysgenic fertility has been in place for around six generations. In the last section of Chapter 10 it was concluded that genotypic intelligence in Western nations has deteriorated by between five and eight IQ points over this time period. Furthermore, the studies showing an inverse relationship between educational level and anticipated fertility among young women in the United States, Britain and The Netherlands in the closing decades of the twentieth century suggest that dysgenic fertility is still present.

In 1940 Frederick Osborn, the president of the American Eugenics Society, proposed what he called "the eugenic hypothesis." This was that dysgenic fertility would shortly come to an end as the knowledge and practice of birth control spread to the lower classes. When this occurred, the professional and middle classes would begin to have more children than the lower classes because they could more easily afford them. Fertility would shortly become eugenic.

Half a century after this hypothesis was advanced, there is no sign that this Panglossian future is being realized. Although the magnitude of dysgenic fertility has declined, it has not disappeared, let alone been superseded by eugenic fertility. This is not surprising, considering that research has shown that in the 1980s and 1990s approximately a third of pregnancies in the United States and Britain are unplanned (Kost and Forrest, 1995; Family Planning Association, 1995). The principal reason for these unplanned pregnancies is that many people continue to use contraception haphazardly and inefficiently. For instance, a study of 1,083 sexually active young males aged 16 and 17 in the United States, carried out in the mid-1980s, found that only 70 percent of them claimed to use contraception and the 30 percent who did not use contraception were the least educated, the least intelligent and the least conscientious (Kirby, Harvey, Cassenius and Novar, 1989). With figures like these, dysgenic fertility is inevitable. Osborn's vision of an imminent future in which everyone uses contraception with perfect efficiency, all births are planned and fertility becomes eugenic was based on a hopelessly optimistic view of human nature. Dysgenic fertility in modern populations remains a problem and is likely to persist indefinitely into the future with steady degradation of the gene pool.

Many people have found it difficult to reconcile the deterioration of genotypic intelligence with the increase in phenotypic intelligence found in many Western nations since the 1930s. We examined this paradox in Chapter 8 and concluded that the rise of phenotypic intelligence has been caused by environmental improvements, largely in nutrition, which have masked the genotypic decline. It is arguable that the important point is that phenotypic intelligence has increased, so the underlying genotypic decline does not matter. Such an argument would be analagous to the "deterioration" of health which has taken place in respect of the loss of genetic immunities against a number of infectious diseases. This is not a cause for concern because the diseases are now controlled. Similarly, it may be said that the deterioration of genotypic intelligence does not matter because we can compensate for it by environmental improvements.

However, the parallel between the control of infectious diseases and the "control" of low intelligence is not exact. It is probable that infectious diseases will continue to be controlled indefinitely, but it is not likely that intelligence will continue to rise indefinitely through environmental improvements. It is much more probable that the environmental improvements will show diminishing returns and eventually cease. When this happens, and if dysgenic fertility continues, phenotypic intelligence will begin to decline, and this will have serious consequences for the quality of life in Western societies.

At least as worrisome is the deterioration of genotypic intelligence, which is taking place in many economically developing nations as they enter the early stages of the demographic transition. These nations need all their resources of human intelligence to develop their economies, find solutions to their social problems and control their population explosions. Whether they will show the same increase in phenotypic intelligence as has been found in the economically developed nations remains to be investigated. Meanwhile, the almost universal dysgenic fertility in these countries, reviewed in Chapter 15, will inevitably impair their progress toward economic and social development.

Turning finally to dysgenic fertility for conscientiousness, the evidence from which to assess the magnitude of the problem is weaker than for intelligence. That some decline has been taking place can be inferred from the inverse associations of fertility with educational attainment and socioeconomic status, because both of these are indirect measures of conscientiousness—from the tendency of criminals to have large numbers of siblings, and from the high fertility of criminals. The high correlations obtained by Tygart (1991) of criminality with number of siblings suggests that genetic deterioration with regard to conscientiousness may be about twice as great as that for intelligence. Our finding that the fertility of criminals in Britain is about 50 percent greater than that of the population as a whole corroborates the conclusion that this is a serious problem. It may well be that dysgenic fertility for conscientiousness and criminality—which has received the least attention from eugenicists, and which has made a significant contribution to the rising crime rates in many Western nations in the second half of the twentieth century—is the most serious of the dysgenic problems confronting modern populations. The time for criminologists to address the issue of dysgenic fertility for crime, and of social scientists more generally to examine dysgenic fertility for conscientiousness, is long overdue.

8. CONCLUSIONS

In the opening chapter of this book we saw how in the 1850s, 1860s and 1870s Benedict Morel in France, and Francis Galton in England, perceived that natural selection was ceasing to operate in Western nations and that this was leading to a deterioration of the genetic quality of the populations in regard to health, intelligence and character. For the next century or so the evidence for this view accumulated, and many biological and social scientists accepted it. Many of them subscribed to the concept of eugenics, the applied science of formulating and implementing policies to counteract genetic deterioration and replace it with genetic improvement.

In the second half of the twentieth century, and increasingly from the 1970s onward, critics asserted that the eugenicists were talking nonsense. They had next to no knowledge of genetics, according to historians like Haller, Ludmerer and Kevles: "They know almost nothing about human inheritance" according to the Galton professor of genetics at University College, London (Jones, 1993, p. 12); their views were "mindless" according to Sir Walter Bodmer (Bodmer and McKie, 1994, p. 236).

In this chapter we have considered the criticisms of the view that the genetic quality of modern populations is deteriorating. These are that there is no genetic determination of intelligence, conscientiousness, crime, educational attainment or socioeconomic status; that there can be an inverse association between intelligence and fertility without genetic deterioration occurring; that there are no genetic differences between the social classes; that there are no such things as bad genes; that the genes for genetic diseases should be preserved, especially in other people, because they make a positive contribution to creative achievement; and that all human types, including the mentally retarded, criminals and psychopaths, are equally valuable. All these arguments have been examined and found wanting. Only one verdict is possible concerning the critics of eugenics who have advanced these arguments, and that is that they have not taken the trouble to examine the research evidence. The eugenicists believed that modern populations are deteriorating genetically. The evidence set out in this book shows they were correct.

The eugenicists thought this a serious problem. We have considered the magnitude of the deterioration and concluded they were right. And they believed that thinking people should consider what policies might be formulated to correct genetic deterioration so that, as Galton originally put it, the failure of natural selection could be compensated by consciously designed selection. In this also they were surely right. The need to counteract genetic deterioration is one of the major components of the case for eugenics. It is not, however, the only component. The eugenicists thought the world would be a better place if people were free of genetic diseases, mental retardation, stupidity, antisocial behavior and crime. This is another thing they were right about. However, I have not been concerned with this component of the eugenic argument in this book, nor have I been concerned with the policies that might be adopted to halt genetic deterioration and improve the genetic quality of human populations. Nevertheless, and contrary to accepted opinion in the closing decades of the twentieth century, these are noble aspirations and I shall take them up in a further volume.

References

Abrams, M. (1985) Demographic correlates of values. In M. Abrams, D. Gerard and N. Timms (eds.), *Values and Social Change in Britain*. Basingstoke: MacMillan.

Alam, I. and Casterline, J.B. (1984) Socio-economic differentials in recent fertility. *World Fertility Survey Comparative Studies, Cross-National Summaries*, No. 33. Voorburg, Netherlands: International Statistical Institute.

Albrecht, S.A., Rosella, J.D. and Patrick, T. (1994) Smoking among low income, pregnant women: Prevalence rates, cessation interventions and clinical implications. *Birth*, 21, 155–162.

Allen, V.L. (1970) Personality correlates of poverty. In V.L. Allen (ed.), *Psychological Factors in Poverty*. Chicago: Markham.

Allgulander, C., Nowak, J. and Rice, J.P. (1991) Psychopathology and treatment of 30,344 twins in Sweden. II. Heritability estimates of psychiatric diagnosis and treatment in 12,884 twin pairs. *Acta Psychiatrica Scandinavia*, 83, 12–15.

Ashurst, H., Balkaran, S. and Casterline, J.B. (1984) Socio-economic differentials in recent fertility. *Comparative Studies*, 42, 1–61.

Atkinson, J. (1958) *Motives in Fantasy, Action and Society*. Princeton, N.J.: Van Nostrand.

Azelmat, M., Ayad, M. and Housni, E.A. (1992) *Enquête nationale sur la population et la santé*. Rabat: Ministère de la Santé Publique.

Babson, S.G. and Phillips, D.S. (1973) Growth and development of twins dissimilar in size at birth. *New England Journal of Medicine*, 289, 937–940.

Bachu, A. (1990) *Fertility of American Women*. Washington, D.C.: U.S. Government Printing Office.

Bachu, A. (1991) *Fertility of American Women: June 1990*. Washington, D.C.: U.S. Government Printing Office.

Bachu, A. (1993) *Fertility of American Women: June 1992*. Washington, D.C.: U.S. Government Printing Office.

Baird, P.A., Anderson, T.W., Newcombe, H.B. and Lowry, R.B. (1988) Genetic disorders in children and young adults: A population study. *American Journal of Human Genetics*, 42, 677–693.

Bajema, C.J. (1963) Estimation of the direction and intensity of natural selection in relation to human intelligence by means of the intrinsic rate of natural increase. *Eugenics Quarterly*, 10, 175–187.

Bajema, C.J. (1968) Relation of fertility to occupational status, IQ, educational attainment and size of family of origin: A follow-up study of the male Kalamazoo public school population. *Eugenics Quarterly*, 15, 198–203.

Bajema, C.J. (1976) *Eugenics: Then and Now*. Stroudsburg, Pa.: Dowden, Hutchinson Ross.

Baker, L.A., Trelvar, S.A., Reynolds, C., Heath, A. and Martin, N.G. (1994) Genetics and educational attainment in Australian twins. *Behavior Genetics*, 24, 138–152.

Barrick, M.R. and Mount, M.K. (1991) The big five personality dimensions and job performance: A meta-analysis. *Personnel Psychology*, 44, 1–26.

Beattie, J.M. (1986) *Crime and the Courts in England 1660–1800*. Oxford: Clarendon Press.

Bell, N.K. (1989) AIDS and women: Remaining ethical issues. *AIDS Education and Prevention*, 1, 22–29.

Belmont, L. and Marolla, F.A. (1973) Birth order, family size and intelligence. *Science*, 182, 1096–1101.

Benjamin, B. (1966) Social and economic differentials in fertility. In J.E. Meade and A.S. Parkes (eds.), *Genetic and Environmental Factors in Human Ability*. Edinburgh: Oliver & Boyd.

Bennett, J.H. (ed.) (1983) *Natural Selection, Heredity and Eugenics*. Oxford: Clarendon Press.

Bennett, T., Braveman, P., Egeter, S. and Kiely, J.L. (1994) Maternal marital status as a risk factor for infant mortality. *Family Planning Perspectives*, 26, 252–256.

Bergeman, C.S., Chipuer, H.M., Plomin, R., Pedersen, N.L., McClean, G.E., Nesselroade, J.R., Costa, P.T. and McCrae, R.R. (1993) Genetic and environmental effects on openness to experience, agreeableness and conscientiousness: An adoption-twin study. *Journal of Personality*, 61, 159–179.

Betzig, L.L. (1986) *Despotism and Differential Reproduction: A Darwinian View of History*. Hawthorne, N.Y.: Aldine.

Blake, J. (1989) *Family Size and Achievement*. Berkeley: University of California Press.

Blaxter, M. (1990) *Health and Lifestyles*. London: Routledge.

Bodmer, W.F. and Cavalli-Sforza, L.L. (1976) *Genetics, Evolution and Man*. San Francisco: Freeman.

Bodmer, W.F. and McKie, R. (1994) *The Book of Man*. London: Little and Brown.

Bohman, M., Cloninger, R., Sigvardsson, S. and von Knorring, A.L. (1987) The genetics of alcohol and related disorders. *Journal of Psychiatric Research*, 21, 447–452.

Bohman, M., Sigvardsson, S. and Cloninger, R.C. (1982) Maternal inheritance of alcohol abuse: Cross-fostering analysis of adopted women. *Archives of General Psychiatry*, 38, 965–969.

Bok, G. (1983) Racism and sexism in Nazi Germany: Motherhood, sterilisation and the state. *Signs*, 8, 413–432.

Boomsma, D.I., Koopmans, J.R., Van Doornen, L.J. and Orlebeke, J.M. (1994) Genetic and social influence on starting to smoke: A study of Dutch adolescent twins and their parents. *Addiction*, 89, 219–226.

Borgerhoff Mulder, M. (1988) Reproductive success in three Kipsigis cohorts. In T.H.

Clutton-Brock (ed.), *Reproductive Success: Studies of Selection and Adaptation in Contrasting Breeding Systems.* Chicago: University of Chicago Press.

Bouchard, T.J. (1993) The genetic architecture of human intelligence. In P.A. Vernon (ed.), *Biological Approaches to the Study of Human Intelligence.* Norwood, N.J.: Ablex.

Bowman, M.L. (1989) Testing individual differences in ancient China. *American Psychologist*, 44, 576–578.

Bradford, E.J.G. (1925) Can present scholastic standards be maintained? *Forum of Education*, 3, 186–194.

Braithwaite, J. (1979) *Inequality, Crime and Public Policy.* London: Routledge and Kegan Paul.

Brand, C. (1995) How many dimensions of personality? The big 5, the gigantic 3 or the comprehensive 6? *Psychologica Belgica*, 34, 257–275.

Breland, H.M. (1974) Birth order, family configuration and verbal achievement. *Child Development*, 45, 1011–1019.

Brody, N. (1992) *Intelligence.* New York: Academic Press.

Broman, S., Nichols, P.L., Shaughnessy, P. and Kennedy, W. (1987) *Retardation in Young Children.* Hillsdale, N.J.: Lawrence Erlbaum.

Burks, B.S. (1928) The relative influence of nature and nurture upon mental development. *Twenty-seventh Yearbook of the National Society for the Study of Education*, 27, 9–38.

Burks, B.S., and Jones, H.E. (1935) A study of differential fertility in two Californian cities. *Human Biology*, 7, 539–554.

Burt, C. (1952) *Intelligence and Fertility*, 2d ed. London: Eugenics Society.

Buss, D. (1994) *The Evolution of Desire.* New York: Basic Books.

Cadoret, R.J., O'Gorman, W., Troughton, E. and Heywood, E. (1985) Alcoholism and anti-social personality. *Archives of General Psychiatry*, 42, 161–167.

Calhoun, C. (1989) Estimating the distribution of desired family size and excess fertility. *Journal of Human Resources*, 24, 709–724.

Calhoun, C. (1991) Desired and excess fertility in Europe and the U.S. *European Journal of Population*, 7, 29–57.

Caplan, N., Choy, M.H. and Whitmore, J.K. (1992) Indochinese refugee families and academic achievement. *Scientific American*, February, 18–24.

Capron, C. and Duyme, L. (1989) Assessment of effects of socio-economic status on IQ in a full cross-fostering design. *Nature*, 340, 552–553.

Carmelli, D., Swan, G.E., Robinette, D. and Fabsitz, R. (1992) Genetic influence on smoking—a study of male twins. *New England Journal of Medicine*, 327, 829–833.

Cattell, R.B. (1937) *The Fight for our National Intelligence.* London: P.S. King.

Cattell, R.B. (1951) The fate of national intelligence: Test of a thirteen year prediction. *Eugenics Review*, 17, 136–148.

Cattell, R.B. (1957) *Personality and Motivation Structure and Measurement.* New York: World Books.

Cattell, R.B. (1971) *Abilities.* New York: Houghton Mifflin.

Cattell, R.B. (1972) *Beyondism.* New York: Pergamon Press.

Cattell, R.B. (1987) *Beyondism*, 2d ed. New York: Praeger.

Cattell, R.B., Schnerger, J.M. and Klein, T.W. (1982) Heritabilities of ego strength, super-ego strength and self-sentiment by multiple abstract variance analysis. *Journal of Clinical Psychology*, 38, 769–779.

Census of Canada (1971) Ottawa: Information Canada.

Census of Canada (1981) Ottawa: Information Canada.

Chagnon, N.A. (1983) *Yanomamo: The Fierce People.* New York: Holt, Rinehart & Winston.

Chamberland, M.E. and Donders, T.J. (1987) Heterosexually acquired infection with the HIV virus. *Annals of Internal Medicine*, 107, 763–766.

Chapman, J.C. and Wiggins, D.M. (1925) The relation of family size to intelligence of offspring and socio-economic status of family. *Pedagogical Seminary and Journal of Genetic Psychology*, 32, 414–421.

Charles, I.E. (1949) *The Changing Size of the Family in Canada.* Ottawa: E. Cloutier.

Churchill, J.H. (1965) The relationship between intelligence and birthweight in twins. *Neurology*, 15, 341–347.

Clark, L.A. and Livesley, W.J. (1994) Two approaches to identifying the dimensions of personality disorder: Convergence on the five factor model. In P.T. Costa and T.A. Widiger (eds.), *Personality Disorders.* Washington, D.C.: American Psychological Association.

Cloninger, C.R., Bohman, M. and Sigvardsson, S. (1981) Inheritance of alcohol abuse: Cross-fostering analysis of adopted men. *Archives of General Psychiatry*, 38, 861–869.

Cloninger, C.R., Christiansen, K.O., Reich, T. and Gottesman, I. (1978) Implications of sex differences in the prevalence of anti-social personality, alcoholism and criminality for family transmission. *Archives of General Psychiatry*, 1978, 35, 941–951.

Coale, A.J. (1986) The decline in fertility in Europe since the eighteenth century as a chapter in human demographic history. In A.J. Coale and S.C. Watkins (eds.), *The Decline of Fertility in Europe.* Princeton, N.J.: Princeton University Press.

Coleman, J.S. (1993) Comment on Preston and Campbell's "Differential fertility and the distribution of traits." *American Journal of Sociology*, 98, 1020–1032.

Coleman, O. and Salt, J. (1992) *The British Population.* Oxford: Oxford University Press.

Conrad, K.M., Flay, B.R. and Hill, D. (1992) Why children start smoking cigarettes: Predictors of onset. *British Journal of Addiction*, 87, 1711–1724.

Cooper, J. (1991) Births outside marriage: Recent trends and associated demographic and social changes. *Population Trends*, 63, 8–18.

Costa, P.T. and McCrae, R.R. (1988) From catalog to classification: Murray's needs and the five factor model. *Journal of Personality and Social Psychology*, 55, 258–265.

Costa, P.T. and McCrae, R.R. (1990) Personality disorders and the five factor personality model of personality. *Journal of Personality Disorders*, 4, 362–371.

Costa, P.T. and McCrae, R.R. (1992) *The NEO personality inventory—revised.* Odessa, Fla.: Psychological Assessment Resources.

Costa, P.T. and Widiger, T.A. (1994) *Personality Disorders and the Five Factor Model of Personality.* Washington D.C.: American Psychological Association.

Cotton, N.S. (1979) The familial incidence of alcoholism: A review. *Journal of Studies in Alcohol*, 40, 89–116.

Crafts, N.F.R. and Mills, T.C. (1994) Trends in real wages in Britain. *Explorations in Economic History*, 31, 195–209.

Crick, F. (1963) Eugenics and genetics. In G. Wolstenholme (ed.), *Man and His Future.* London: Churchill.

Crow, J.F. (1966) The quality of people: Human evolutionary changes. *Bio Science*, 16, 863–867.

Crow, J.F. (1970) Do genetic factors contribute to poverty? In V.L. Allen (ed.), *Psychological Factors in Poverty.* Chicago: Markham.

Crowe, R. (1972) The adopted offspring of women criminal offenders. *Archives of General Psychiatry*, 27, 600–603.

Crowe, R. (1975) An adoptive study of psychopathy. In R. Fieve, D. Rosenthal and H. Brill (eds.), *Genetic Research in Psychiatry.* Baltimore: Johns Hopkins University Press.

Daly, M. and Wilson, M. (1983) *Sex, Evolution and Behavior.* Boston: Willard Grant Press.

Damrin, D.E. (1949) Family size and sibling age, sex and position as related to certain aspects of adjustment. *Journal of Social Psychology*, 29, 93–102.

Darwin, C. (1871) *The Descent of Man and Selection in Relation to Sex.* London: MacMillan.

Darwin, L. (1926) *The Need for Eugenic Reform.* London: Murray.

Davenport, C. (1911) *Heredity in Relation to Eugenics.* New York: Henry Holt.

Degler, C.N. (1991) *In Search of Human Nature.* Oxford: Oxford University Press.

Desplanques, G. (1988) Comportements demographiques: Une fecondite maitrisee. *Sociologie du Travail*, 38, 353–365.

Dickemann, M. (1979) The ecology of mating systems in hypergynous dowry societies. *Social Science Information*, 18, 163–195.

Dill, S. (1898) *Roman Society in the Last Century of the Western Empire.* London: MacMillan.

Dobzhansky, T. (1962) *Mankind Evolving.* New Haven: Yale University Press.

Douglas, J.W.B. (1964) *The Home and the School.* London: MacGibbon and Kee.

Douglas, J.W.B., Ross, J.M., Hammond, W.A. and Mulligan, D.G. (1966) Delinquency and social class. *British Journal of Delinquency*, 6, 294–302.

Driesen, I. (1972) Some observations on the family unit, relatedness, and the practice of polygyny in the Ife division of Western Nigeria. *Africa*, 47, 44–56.

Duff, J.F. and Thomson, G.H. (1923) The social and geographical distribution of intelligence in Northumberland. *British Journal of Psychology*, 14, 192–198.

Duncan, O.D. (1952) Is the intelligence of the general population declining? *American Sociological Review*, 17, 401–407.

Dupont, A. (1989) 140 years of Danish studies on the prevalence of mental retardation. *Acta Psychiatrica Scandinavia*, 79 (suppl. 348), 105–112.

Eaves, L.J., Eysenck, H.J. and Martin, N.G. (1989) *Genes, Culture and Personality.* London: Academic Press.

Elder, G. (1968) Achievement motivation and intelligence in occupational mobility. *Sociometry*, 31, 327–354.

Elley, W.B. (1969) Changes in mental ability in New Zealand school children. *New Zealand Journal of Educational Studies*, 4, 140–155.

Elliott, D.S. and Huizinga, D. (1983) Social class and delinquent behavior in a national youth pattern. *Criminology*, 21, 149–177.

Ellis, L. (1988) The victimful-victimless crime distinction, and seven universal demographic correlates of victimful criminal behavior. *Personality and Individual Differences*, 9, 525–548.

Emmett, W.G. (1950) The trend of intelligence in certain districts of England. *Population Studies*, 3, 324–337.

Eysenck, H.J. (1976) *Sex and Personality*. London: Open Books.

Eysenck, H.J. (1979) *The Structure and Measurement of Intelligence*. London: Springer-Verlag.

Eysenck, H.J. (1992) The definition and measurement of psychoticism. *Personality and Individual Differences*, 13, 757–785.

Eysenck, H.J. and Gudjonsson, G.H. (1989) *The Causes and Cures of Criminality*. New York: Plenum Press.

Falconer, D.S. (1960) *An Introduction to Quantitative Genetics*. New York: Ronald Press.

Family Planning Association. (1995) *Report on Costs and Benefits of Family Planning*. London: Family Planning Association.

Farrington, D.P. and West, D.J. (1990) The Cambridge Study in Delinquent Development: A long term follow-up of 411 London males. In H.J. Kerner and G. Kaiser (eds.), *Criminality: Personality, Behavior and Life History*. Berlin: Springer-Verlag.

Fisher, R.A. (1918) The correlation between relatives on the supposition of Mendelian inheritance. *Transactions of the Royal Society of Edinburgh*, 52, 399–433.

Fisher, R.A. (1929) *The Genetical Theory of Natural Selection*. Oxford: Clarendon Press.

Fishler, K., Azen, C.G., Friedman, E.G. and Koch, R. (1989) School achievement in treated PKU children. *Journal of Mental Deficiency Research*, 33, 493–498.

Flay, B.R., d'Avernas, J.R., Best, J.A., Kersell, M.W. and Ryan, K.B. (1983) In P.J. McGrath and P. Firestone (eds.), *Pediatric and Adolescent Behavioral Medicine: Issues in Treatment*. New York: Springer.

Flynn, J.R. (1984) The mean IQ of Americans: Massive gains 1932 to 1978. *Psychological Bulletin*, 95, 29–51.

Flynn, J.R. (1987) Massive IQ gains in 14 nations: What IQ tests really measure. *Psychological Bulletin*, 171–191.

Folling, I.A. (1934) Uber ausscheidung von phenylbrenztraubensaure in den harn als stoffwechselanomie in verbindung mit imbezilitat. *Zeitschrift fur physiologisch chemie*, 227, 169–176.

Forrest, J.D. and Singh, S. (1990) The sexual and reproductive behavior of American women, 1982–1988. *Family Planning Perspectives*, 22, 206–214.

Freedman, R., Whelpton, P.K. and Campbell, A.A. (1959) *Family Planning, Sterility and Population Growth*. New York: McGraw-Hill.

Frisancho, A.R., Matos, J. and Flegel, P. (1983) Maternal nutritional status and adolescent pregnancy outcome. *American Journal of Clinical Nutrition*, 38, 739–746.

Frisch, R. (1978) Population, food intake and fertility. *Science*, 199, 22–30.

Frisch, R. (1984) Body fat, puberty and fertility. *Biology Review*, 59, 161–188.

Fujikura, T. and Froehlich, L.A. (1974) Mental and motor development in monozygotic co-twins with dissimilar birthweights. *Pediatrics*, 53, 884–889.

Furnham, A. (1990) *The Protestant Work Ethic*. London: Routledge.

Galbraith, R.C. (1982) Sibling spacing and intellectual development: A closer look at the confluence model. *Developmental Psychology*, 18, 151–173.

Galton, F. (1865) Hereditary talent and character. *MacMillan's Magazine*, 12, 157–166; 318–327.

Galton, F. (1869) *Hereditary Genius.* London: MacMillan.
Galton, F. (1873) Hereditary improvement. *Frazer's Magazine*, 7, 116–130.
Galton, F. (1874) *English Men of Science: Their Nature and Nurture.* London: Methuen.
Galton, F. (1883) *Inquiries into Human Faculty and its Development.* London: Dent.
Galton, F. (1901) The possible improvement of the human breed under the existing conditions of law and sentiment. *Nature*, 64, 659–665.
Galton, F. (1908) *Memories of My Life.* London: Methuen.
Galton, F. (1908) Local associations for promoting eugenics. *Nature*, 78, 645–647.
Galton, F. (1909) *Essays in Eugenics.* London: Eugenics Society.
Galton, F. (1910) *The Eugenic College of Kantsaywhere.* Unpublished extracts in K. Pearson's *Life of Francis Galton.*
General Household Survey. (1989) *Report.* London: HMSO.
General Household Survey. (1993) *Report.* London: HMSO.
Gerard, D. (1985) Religious attitudes and values. In M. Abrams, D. Gerard and N. Timms (eds.), *Values and Social Change in Britain.* Basingstoke: MacMillan.
Gilbert, P. (1993) *The A–Z Reference Book of Syndromes and Inherited Disorders.* London: Chapman and Hall.
Giles-Bernadelli, B.M. (1950) The decline of intelligence in New Zealand. *Population Studies*, 4, 200–208.
Glass, D.V. (1967) Fertility trends in Europe since the second world war. In *Proceedings of the University of Michigan Conference on Fertility and Family Planning.* Ann Arbor: University of Michigan.
Glendinning, A., Shucksmith, J. and Hendry, L. (1994) Social class and adolescent smoking behavior. *Social Science and Medicine*, 38, 1449–1460.
Goldblatt, P. (1990) Social class mortality differences. In C.G.N. Mascie-Taylor (ed.), *Biosocial Aspects of Social Class.* Oxford: Oxford University Press.
Goldthorpe, J.H., Llewellyn, C. and Payne, C. (1987) *Social Mobility and Class Structure in Modern Britain.* Oxford: Clarendon Press.
Goldthorpe, J.H., Lockwood, D., Bechhofer, F. and Platt, J. (1969) *The Affluent Worker.* Cambridge: Cambridge University Press.
Gorer, G. (1971) *Sex and Marriage in England Today.* London: Nelson.
Gottesman, I.I. (1963) Heritability of personality: A demonstration. *Psychological Monographs*, 77, No. 572.
Gottesman, I.I. and Goldsmith, H.H. (1995) Developmental psychopathology of antisocial behavior: Inserting genes into its ontogenesis and epigenesis. In C. Nelson (ed.), *Threats to Optimal Development.* Hillsdale, N.J.: Lawrence Erlbaum.
Gould, S.J. (1976) Biological potential versus biological determinism. *Natural History*, 85, 345–349.
Gould, S.J. (1984) Singapore's patrimony and matrimony. *Natural History*, 93, 22–29.
Graham, R.K. (1970) *The Future of Man.* North Quincy, Mass.: Christopher.
Graham, R.K. (1987) Combating dysgenic trends in modern society: The repository for germinal choice. *Mankind Quarterly*, 27, 327–335.
Grindstaff, C.F., Balakrishnan, T.R. and Dewit, D.J. (1991) Educational attainment, age at first birth and lifetime fertility: An analysis of Canadian fertility survey data. *Canadian Review of Sociology and Anthropology*, 28, 324–339.
Grove, V. (1991) To life's frontier with the doctor in genes. *Sunday Times*, 17 November, 2.3.
Gruenwald, P., Funakawa, H., Mitani, S., Nishimura, T. and Takeuchi, S. (1967) Influ-

ence of environmental factors on foetal growth in man. *Lancet*, 13 May, 1026–1028.

Haines, M.R. (1989) Social class differentials during fertility decline: England and Wales revisited. *Population Studies*, 43, 305–323.

Haines, M.R. (1992) Occupation and social class during fertility decline: Historical perspectives. In J.R. Gillis, L.A. Tilly and D. Levine (eds.), *The European Experience of Declining Fertility*. Oxford: Blackwell.

Haller, M. (1963) *Eugenics*. New Brunswick, N.J.: Rutgers University Press.

Harpur, T.J., Hart, S.D. and Hare, R.D. (1994) *Personality of the Psychopath*. In P.T. Costa and T.A. Widiger (eds.), *Personality Disorders*. Washington, D.C.: American Psychological Association.

Harrell, T.W. and Harrell, M.S. (1945) Army General Classification Test scores for civilian occupations. *Educational and Psychological Measurement*, 5, 229–239.

Harris, J. (1992) *Wonderwoman and Superman*. Oxford: Oxford University Press.

Harris, M. and Ross, E.B. (1987) *Death, Sex and Fertility*. New York: Columbia University Press.

Hassan, J. (1989) Way of life, stress and differences in morbidity between occupational classes. In J. Fox (ed.), *Health Inequalities in European Countries*. Aldershot: Gower.

Haviland, W. (1967) Statue at Tikal: Implications for ancient Maya demography and social organisation. *American Antiquity*, 32, 326–335.

Hayami, A. (1980) Class differences in marriage and fertility among Tokugawa villagers in Mino province. *Keio Economic Studies*, 17, 1–16.

Hayman, R.L. (1990) Presumptions of justice: Law, politics and the mentally retarded parent. *Harvard Law Review*, 103, 1202–1271.

Heath, A.C. (1981) *Social Mobility*. Glasgow: Fontana.

Heath, A.C., Berg K., Eaves, L.J., Solaas, M.H., Corey, L.A., Sundet, J., Magnus, P. and Nancy, W.E. (1985) Educational policy and the heritability of educational attainment. *Nature*, 314, 734–736.

Heath, A.C. and Martin, N.G. (1993) Genetic models for the natural history of smoking: Evidence for a genetic influence on smoking persistence. *Addictive Behavior*, 18, 19–34.

Heath, A.C., Meyer, J., Jardine, R., and Martin, N.G. (1991) The inheritance of alcohol consumption patterns in a general population twin sample: Determinants of consumption frequency and quantity consumed. *Journal of Studies on Alcohol*, 52, 425–433.

Hed, H. (1987) Trends in opportunity for natural selection in the Swedish population during the period 1650–1980. *Human Biology*, 59, 785–797.

Hendrichsen, L., Skinhoj, K. and Andersen, C.E. (1986) Delayed growth and reduced intelligence in 9–17 year old intra-uterine growth retarded children compared with their monozygous co-twins. *Acta Paediatrica Scandinavia*, 75, 31–35.

Herlihy, D. (1965) Population, plague and social change in rural Pistoia. *Economic History Review*, 2, 225–244.

Heron, D. (1906) *On the Relation of Fertility in Man to Social Status*. London: Dulan.

Herrman, L. and Hogben, L. (1932) The intellectual resemblance of twins. *Proceedings of the Royal Society of Edinburgh*, 53, 105–129.

Herrnstein, R.J. (1971) *IQ in the Meritocracy*. Boston: Little, Brown.

Herrnstein, R.J. and Murray, C. (1994) *The Bell Curve*. New York: Free Press.

Higgins, J.V., Reed, E.W. and Reed, S.G. (1962) Intelligence and family size: A paradox resolved. *Eugenics Quarterly*, 9, 84–90.

Hill, K. and Kaplan, H. (1988) Trade-offs in male and female reproductive strategies among the Ache. In L. Betzig, M. Burgerhoff Mulder and P. Turke (eds.), *Human Reproductive Behavior.* Cambridge: Cambridge University Press.

Himmelfarb, G. (1995) *The De-moralisation of Society.* London: Institute of Economic Affairs.

Hirschberg, N. and Itkin, S. (1978) Graduate student success in psychology. *American Psychologist*, 33, 1083–1093.

Ho, D.Y.F. (1979) Sibship variables as determinants of intellectual-academic ability in Hong Kong pupils. *Genetic Psychology Monographs*, 100, 21–39.

Ho, P. (1959) Aspects of social mobility in China, 1368–1911. *Comparative Studies in Sociology and History*, 1, 330–359.

Hogben, L. (1931) *Genetic Principles in Medicine and Social Science.* London: Williams and Norgate.

Hogben, L. (1938) *Science for the Citizen.* London: Allen and Unwin.

Hopper, J.L., White, V.M., Macaskill, G.T., Hill, D.J. and Clifford, C.A. (1992) Alcohol use, smoking habits and the Junior Eysenck Personality Questionnaire in adolescent Australian twins. *Acta Genetica Medica Gemelloli*, 41, 311–324.

Howell, N. (1979) *Demography of the Dobe !Kung.* New York: Academic Press.

Hunter, J.E. and Hunter, R.F. (1984) Validity and utility of alternative predictors of job performance. *Psychological Bulletin*, 96, 72–98.

Husen, T. (1959) *Psychological Twin Research: A Methodological Study.* Uppsala, Sweden: Almquist and Wiksells.

Hutchings, B. and Mednick, S.A. (1977) Criminality in adoptees and their adoptive and biological parents: A pilot study. In S.A. Mednick and K.O. Christiansen (eds.), *Biosocial Bases of Criminal Behavior.* New York: Gardner Press.

Huxley, J.S. (1936) Eugenics and Society. *Eugenics Review*, 28, 11–31.

Huxley, J.S. (1941) *Man Stands Alone.* London: Harper.

Huxley, J.S. (1942) *Evolution, the Modern Synthesis.* London: Allen and Unwin.

Huxley, J.S. (1962) Eugenics in evolutionary perspective. *Eugenics Review*, 54, 123–124.

Inge, W.R. (1927) *Outspoken Essays.* London: Longmans, Green.

Irons, W. (1979) Cultural and biologic success. In N.A. Chagnon and W. Irons (eds.), *Evolutionary Biology and Human Social Behavior: An Anthropological Perspective.* North Scituate, Mass.: Duxbury Press.

Jahoda, M. (1989) A helping hand to evolution? *Contemporary Psychology*, 34, 816–817.

Jarvelin, M-R., Laara, E., Rantakallio, P., Moilanen, I. and Isohanni, M. (1994) Juvenile delinquency, education and mental disability. *Exceptional Children*, 61, 230–241.

Jencks, C.R. (1972) *Inequality.* New York: Basic Books.

Jencks, C.R. (1979) *Who Gets Ahead?* New York: Basic Books.

Jencks, C.R. (1992) *Re-thinking Social Policy.* Cambridge: Harvard University Press.

Jensen, A.R. (1972) *Genetics and Education.* London: Methuen.

Jensen, A.R. (1980) *Bias in Mental Testing.* London: Methuen.

Jensen, A.R. (1983) Effects of inbreeding on mental ability factors. *Personality and Individual Differences*, 4, 71–88.

Jensen, A.R. and Reynolds, C.R. (1982) Race, social class and ability patterns on the WISC-R. *Personality and Individual Differences*, 3, 423–438.

Johnson, D.M. (1948) Applications of the standard score IQ to social statistics. *Journal of Social Psychology*, 27, 217–227.

Jones, D.C. and Carr-Saunders, A.M. (1927) The relation between intelligence and social status among orphan children. *British Journal of Psychology*, 17, 343–364.

Jones, S. (1993) *The Language of the Genes.* London: Harper Collins.

Kaelber, C.T. and Pugh, T.F. (1969) Influence of intra-uterine relations on the intelligence of twins. *New England Journal of Medicine*, 280, 1030–1034.

Kaelble, H. (1985) *Social Mobility in the 19th and 20th Centuries.* Leamington Spa: Berg.

Kamin, L.J. (1974) *The Science and Politics of IQ.* London: Penguin Books.

Kannae, K. and Pendleton, B.F. (1994) Fertility attitudes among male Ghanaian government employees. *Journal of Asian and African Studies*, 29, 65–76.

Kass, L.R. (1972) New beginnings in life. In M.P. Hamilton (ed.), *The New Genetics and the Future of Man.* New York: W.B. Eerdmans.

Kaufman, A.S. and Doppelt, J.E. (1976) Analysis of WISC-R standardisation data in terms of the stratification variables. *Child Development*, 47, 165–171.

Kaye, H.L. (1987) *The Social Meaning of Modern Biology.* New Haven: Yale University Press.

Keller, L.M., Bouchard, T.J., Arvey, R.D., Segal, N.L. and Davis, R.V. (1992) Work values: Genetic and environmental influences. *Journal of Applied Psychology*, 77, 79–88.

Kevles, D.J. (1985) *In the Name of Eugenics.* New York: A.A. Knopf.

Keynes, M. (1993) Sir Francis Galton. In M. Keynes (ed.), *Sir Francis Galton: The Legacy of His Ideas.* London: MacMillan.

Kiernan, K.E. (1989) Who remains childless? *Journal of Biosocial Science*, 21, 387–398.

Kiernan, K.E. and Diamond, I. (1982) Family of origin and educational influences on age at first birth: The experiences of a British birth cohort. London: Centre for Population Studies, Research Paper No. 81.

Kirby, D., Harvey, P.D., Cassenius, D. and Novar, M. (1989) A direct mailing to teenage males about condom use. *Family Planning Perspectives*, 21, 12–18.

Kiser, C.V. (1970) Changing patterns of fertility in the United States. *Social Biology*, 17, 302–315.

Kiser, C.V. and Frank, M.E. (1967) Factors associated with the low fertility of nonwhite women of college attainment. *Milbank Memorial Fund Quarterly*, 15, 427–449.

Knodel, J. and van de Walle, E. (1986) Lessons from the past: Policy implications of historical fertility studies. In A.J. Coale and S.C. Watkins (eds.), *The Decline of Fertility in Europe.* Princeton, N.J.: Princeton University Press.

Kohn, M.I. and Schooler, C. (1969) Class, occupation and orientation. *American Sociological Review*, 34, 659–678.

Kolvin, I., Miller, F.J.W., Scott, D.M., Gatzanis, S.R.M. and Fleeting, M. (1990) *Continuities of Deprivation.* Aldershot: Avebury.

Koopmans, J.R. and Boomsma, D.I. (1993) Bivariate genetic analysis of the relation between alcohol and tobacco use in adolescent twins. *Psychiatric Genetics*, 3, 172–186.

Kopp, M. (1936) Legal and medical aspects of eugenic sterilisation in Germany. *American Sociological Review*, 1, 766–770.

Kost, K. and Forrest, J.D. (1995) Intention status of U.S. births in 1988: Differences by mothers' socio-economic and demographic characteristics. *Family Planning Perspectives*, 27, 11–17.

Kravdal, O. (1992) The emergence of a positive relation between education and third birth rates in Norway with supportive evidence from the United States. *Population Studies*, 46, 459–475.

Krueger, R.F., Schmutte, P.S., Caspi, A., Moffit, T.E., Campbell, K. and Silva, P.A. (1994) Personality traits are linked to crime among men and women: Evidence from a birth cohort. *Journal of Abnormal Psychology*, 103, 328–338.

Kunst, A.E. and Mackenbach, J.P. (1994) The size of mortality differences associated with educational level in nine industrialised countries. *American Journal of Public Health*, 84, 932–937.

Lange, J. (1929) *Crime as Destiny*. Leipzig: Thieme.

Lasker, G.W. (1991) Introduction. In C.G.N. Mascie-Taylor and G.W. Lasker (eds.), *Applications of Biological Anthropology to Human Affairs*. Cambridge: Cambridge University Press.

Lawrence, E.M. (1931) An investigation into the relation between intelligence and inheritance. *British Journal of Psychology*, Monograph Supplement, 16, 1–80.

Leahy, A. (1935) Nature-nurture and intelligence. *Genetic Psychology Monographs*, 17, 4.

Leclerc, A. (1989) Differential mortality by cause of death: Comparison between selected European countries. In J. Fox (ed.), *Health Inequalities in European Countries*. Aldershot: Gower.

Leclerc, A., Lert, F. and Fabien, C. (1990) Differential mortality: Some comparisons between England and Wales, Finland and France, based on inequality measures. *International Journal of Epidemiology*, 4, 1–10.

Lederberg, J. (1963) Biological future of man. In G. Wolstenholme (ed.), *Man and His Future*. London: Churchill.

Lentz, T.F. (1927) The relation of IQ to size of family. *Journal of Educational Psychology*, 18, 486–496.

Lewis, O. (1961) *The Children of Sanchez*. New York: Random House.

Lewontin, R.C. (1993) *The Doctrine of DNA*. London: Penguin Books.

Lewontin, R.C., Rose, S. and Kamin, L. (1989) *Not in Our Genes*. New York: Pantheon.

Lichtenstein, P., Herschberger, S.L. and Pedersen, N.L. (1995) Dimensions of occupations: Genetic and environmental influences. *Journal of Biosocial Science*, 27, 193–206.

Livesley, W.J., Jang, K.L., Jackson, D.N. and Vernon, P.A. (1993) Genetic and environmental contributions to dimensions of personality disorder. *American Journal of Psychiatry*, 150, 1826–1831.

Loehlin, J.C. and Nichols, R.C. (1976) *Heredity, Environment and Personality*. Austin: University of Texas Press.

Loehlin, J.C., Willerman, L. and Horn, J.M. (1987) Personality resemblance in adoptive families: A 10 year follow up. *Journal of Personality and Social Psychology*, 53, 961–969.

Lorr, M. (1986) *Interpersonal Style Inventory Manual*. Los Angeles: Western Psychological Services.

Ludmerer, K. (1972) *Genetics and American Society*. Baltimore: Johns Hopkins University Press.

Lumey, L.H. and van Poppel, F.W. (1994) The Dutch famine of 1944–45: Mortality and morbidity in past and present generations. *Social History of Medicine*, 7, 229–246.

Luster, T. and Small, S.A. (1994) Factors associated with sexual risk-taking behaviors among adolescents. *Journal of Marriage and the Family*, 56, 622–632.

Lynn, R. (1977) The intelligence of the Chinese and Malays in Singapore. *Mankind Quarterly*, 18, 125–128.

Lynn, R. (1990) The role of nutrition in secular increases in intelligence. *Personality and Individual Differences*, 11, 273–288.

Lynn, R. (1991) *The Secret of the Miracle Economy.* London: Social Affairs Unit.

Lynn, R. (1995) Dysgenic fertility for crime. *Journal of Biosocial Science*, 27, 405–408.

Lynn, R., Devane, S. and O'Neill, B. (1984) Extending the boundaries of psychoticism: health care and the self-sentiment. *Personality and Individual Differences*, 5, 397–402.

Lynn, R. and Hampson, S. (1986) The rise of national intelligence: Evidence from Britain, Japan and the U.S.A. *Personality and Individual Differences*, 7, 23–32.

Lynn, R., Hampson, S. and Howden, V. (1988) The intelligence of Scottish children, 1932–1986. *Studies in Education*, 6, 19–25.

Lynn, R., Hampson, S. and Mullineux, J.C. (1987) A long term increase in the fluid intelligence of British children. *Nature*, 328, 797.

Lynn, R. and Pagliari, C. (1994) The intelligence of American children is still rising. *Journal of Biosocial Science*, 65–67.

Mackintosh, N.J. (1975) Review of the science and politics of IQ. *Quarterly Journal of Experimental Psychology*, 27, 272–286.

Mackintosh, N.J. and Mascie-Taylor, C.G.N. (1984) The IQ question. In *Education for All.* Cmnd paper 4453. London: HMSO.

Maguire, M. (1994) Crime, statistics, patterns and trends. In M. Maguire, R. Morgan and R. Reiner (eds.), *The Oxford Handbook of Criminology.* Oxford: Clarendon Press.

Maguire, M., Morgan, R. and Reiner, R. (eds.) (1994) *The Oxford Handbook of Criminology.* Oxford: Clarendon Press.

Mann, M. (1986) Work and the work ethic. In R. Jowell, S. Witherspoon and L. Brook (eds.), *British Social Attitudes.* Aldershot: Gower.

Martin, E.D. and Sher, K.J. (1994) Family history of alcoholism, alcohol use disorders and the five factor model of personality. *Journal of Studies on Alcohol*, 55, 81–90.

Martin, N. and Jardine, R. (1986) Eysenck's contributions to behavior genetics. In S. Modgil and C. Modgil (eds.), *Hans Eysenck: Consensus and Controversy.* London: Falmer Press.

Martin, T.C. and Juárez, F. (1995) The impact of women's education on fertility in Latin America. *Family Planning Perspectives*, 21, 52–57.

Mascie-Taylor, C.G.N. (1990) The biology of social class. In C.G.N. Mascie-Taylor (ed.), *Biological Aspects of Social Class.* Oxford: Oxford University Press.

Mason, D.A. and Frick, P.J. (1994) The heritability of anti-social behavior. *Journal of Psychopathology and Behavioral Assessment.* 16, 301–323.

Mason, J.O., Noble, H.K., Lindsey, H.K., Kolbe, P.L.J., Van Ness, P., Bowen, F.S., Drotman, D.P. and Rosenberg, M.L. (1987) Current CDS efforts to prevent and control HIV infection and AIDS in the United States through information and education. *Public Health Reports*, 103, 255–260.

Matthews, K., Kelsey, S. Meilahn, F., Kuller, L. and Wing, R. (1989) Educational attainment and behavioral and biological risk factors for coronary heart disease in middle-aged women. *American Journal of Epidemiology*, 129, 1132–1144.

Matthews, K.A., Batson, C.D., Horn, J. and Rosenman, R.H. (1981) "Principles in his nature which interest him in the fortune of others": The heritability of empathic concern for others. *Journal of Personality*, 49, 237–247.

Maxwell, J. (1969) Intelligence, education and fertility: A comparison between the 1932 and 1947 Scottish surveys. *Journal of Biosocial Science*, 1, 247–271.

McClelland, D. (1961) *The Achieving Society.* New York: Free Press.

McKay, D. (1963) Eugenics and genetics. In G. Wolstenholme (ed.), *Man and His Future.* London: Churchill.

McKusick, V.A. (1992) *Mendelian Inheritance in Man.* Baltimore: Johns Hopkins University Press.

McNeill, W.H. (1977) *Plagues and Peoples.* Oxford: Oxford University Press.

McNemar, Q. (1942) *The Revision of the Stanford-Binet Scale.* Boston: Houghton Mifflin.

Mead, M. (1935) *Sex and Temperament in Three Primitive Societies.* New York: Morrow.

Mealey, L. (1985) The relationship between social status and biological success: A case study of the Mormon religious hierarchy. *Ethology and Sociobiology*, 6, 249–257.

Mednick, S.A., Gabrielli, W.F. and Hutchings, B. (1984) Genetic influences in criminal convictions: Evidence from an adoption cohort. *Science*, 224, 891–894.

Mercer, A.J. (1985) Smallpox and the epidemiological-demographic change in Europe: The role of vaccination. *Population Studies*, 39, 287–307.

Miller, A.K. and Corsellis, J.A.N. (1977) Evidence for a secular increase in human brain weight during the past century. *Annals of Human Biology*, 4, 253–257.

Miller, B.C. and Sneesby, K.R. (1988) Educational correlates of adolescents' sexual attitudes and behavior. *Journal of Youth and Adolescence*, 17, 521–530.

Mischel, W. (1958) Preference for delayed reinforcement: An experimental study of a cultural observation. *Journal of Abnormal and Social Psychology*, 56, 57–61.

Mischel, W. (1961) Father absence and delay of gratification. *Journal of Abnormal and Social Psychology*, 63, 116–124.

Moatti, J.P., Bajos, N., Durbec, J.P., Menard, C. and Serrand, C. (1991) Determinants of condom use among French heterosexuals with multiple partners. *American Journal of Public Health*, 81, 106–109.

Modell, B. and Kuliev, A.M. (1989) Impact of public health on human genetics. *Clinical Genetics*, 36, 286–298.

Moffitt, R. (1992) Incentive effects of the U.S. welfare system: A review. *Journal of Economic Literature*, 30, 1–61.

Moran, E. (1979) *Human Adaptability: An Introduction to Ecological Anthropology.* Belmont, Calif.: Duxbury Press.

Morel, B.A. (1857) *Traité des dégenérescenses physiques, intellectuelles et morales de l'espèce humaine.* Paris: Larouse.

Morgan, P. (1995) *Farewell to the Family.* London: IEA Health and Welfare Unit.

Morrison, A.S., Kirschner, J. and Molha, A. (1977) Life cycle events in 15th century Florence: Records of the Monte Delle Doti. *American Journal of Epidemiology*, 106, 487–492.

Moshinsky, P. (1939) The correlation between fertility and intelligence within social classes. *Sociological Review*, 31, 144–165.

Muller, H.J. (1935) *Out of the Night.* New York: Vanguard Press.

Muller, H.J. (1950) Our load of mutations. *American Journal of Human Genetics*, 45, 142–175.

Muller, H.J. (1963) Genetic progress by voluntarily conducted germinal choice. In G. Wolstenholme (ed.), *Man and His Future.* London: Churchill.

Muller, H.J. et al. (1939) The Geneticists' Manifesto. *Eugenical News*, 24, 63–64.

Munsinger, H. (1975) Children's resemblance to their biological and adopting parents in two ethnic groups. *Behavior Genetics*, 5, 239–254.

Murdock, G.P. (1967) *Ethnographic Atlas.* Pittsburgh, Pa.: University of Pittsburgh Press.

Murray, C. (1984) *Losing Ground.* New York: Basic Books.

Murray, C. (1994) *Underclass: The Crisis Deepens.* London: IEA Health and Welfare Unit.

Neel, J.V. (1983) Some base lines for human evolution and the genetic implications of recent cultural developments. In D.J. Ortner (ed.), *How Humans Adapt.* Washington, D.C.: Smithsonian Institution.

Neel, J.V. (1994) *Physician to the Gene Pool.* New York: J. Wiley.

Newman, H.H., Freeman, F.N. and Holzinger, K.J. (1937) *Twins: A Study of Heredity and Environment.* Chicago: University of Chicago Press.

Nisbet, J.D. (1958) Intelligence and family size, 1949–56. *Eugenics Review*, 49, 4–5.

Nohara-Atoh, M. (1980) *Social Determinants of Reproductive Behavior in Japan.* Ph.D. Thesis, University of Michigan, Ann Arbor, Michigan.

Norman, W.T. (1963) Toward an adequate taxonomy of personality attributes: Replicated factor structure in peer nomination personality ratings. *Journal of Abnormal and Social Psychology*, 66, 574–583.

Nystrom, S., Bygren, L.O. and Vining, D.R. (1991) Reproduction and level of intelligence. *Scandinavian Journal of Social Medicine*, 19, 187–189.

O'Donnell, L., Doval, A.S., Duran, R. and O'Donnell, C.R. (1995) Predictors of condom acquisition after an STD clinic visit. *Family Planning Perspectives*, 27, 29–33.

Office of Population, Censuses and Surveys (1987) *Report on Voluntary Work.* London: HMSO.

Office of Population, Censuses and Surveys (1990) *Birth Statistics.* London: HMSO.

Ogawa, N. and Retherford, R.D. (1993) The resumption of fertility decline in Japan: 1973–92. *Population and Development Review*, 19, 703–741.

Olivier, G. and Devigne, G. (1983) Biology and social structure. *Journal of Biosocial Science*, 15, 379–389.

Olsen, J. and Frische, G. (1993) Social differences in reproductive health. *Scandinavian Journal of Social Medicine*, 21, 90–97.

Osborn, F. (1940) *Preface to Eugenics.* New York: Harper.

Osborn, F. (1951) *Preface to Eugenics*, 2d ed. New York: Harper.

Osborn, F. and Bajema, C. (1972) The eugenic hypothesis. *Social Biology*, 19, 337–345.

Osborne, R.T. (1975) Fertility, IQ and school achievement. *Psychological Reports*, 37, 1067–1073.

O'Toole, B.I. and Stankov, L. (1992) The ultimate validity of psychological tests. *Personality and Individual Differences*, 13, 699–716.

Papavassiliou, I.T. (1954) Intelligence and family size. *Population Studies*, 7, 222–226.

Pappas, G., Queen, S., Hadden, W. and Fisher, G. (1993) The increasing disparity in mortality between socio-economic groups in the United States. *New England Journal of Medicine*, 329, 103–109.

Pauli, W.F. (1949) *The World of Life.* Boston: Houghton Mifflin.

Payling, S.J. (1992) Social mobility, demographic change, and landed society in late medieval England. *Economic History Review*, 45, 51–73.

Pearson, K. (1901) *National Life from the Standpoint of Science.* London: Methuen.

Pearson, K. (1903) On the inheritance of the mental and moral characters in man. *Journal of the Anthropological Institute of Great Britain and Ireland*, 33, 179–237.

Pearson, K. (1912) *The Groundwork of Eugenics.* Cambridge: Eugenics Laboratory.

Pearson, R. (1992) *Shockley on Eugenics and Race.* Washington, D.C.: Scott Townsend.

Pedersen, N.L., Friberg, L., Floderus-Myrhed, B., McClean, G.E. and Plomin, R. (1984) Swedish early separated twins: Identification and characterisation. *Acta Geneticae Medicae et Gemellologiae*, 33, 243–250.

Penrose, L.S. (1948) The supposed threat of declining intelligence. *American Journal of Mental Deficiency*, 58, 114–118.

Penrose, L.S. (1950) Genetical influences on the intelligence level of the population. *British Journal of Psychology*, 40, 128–136.

Penrose, L.S. (1967) Presidential address—the influence of the English tradition in human genetics. *Proceedings of the Third International Congress of Human Genetics.* Baltimore: Johns Hopkins University Press.

Petrill, S.A. and Thompson, L.A. (1994) The effects of gender upon heritability and common environmental estimates in measures of scholastic achievement. *Personality and Individual Differences*, 16, 631–640.

Phillips, D. and Harding, S. (1985) The structure of moral values. In M. Abrams, D. Gerard and N. Timms (eds.), *Values and Social Change in Britain.* Basingstoke: MacMillan.

Plomin, R. (1986) *Development, Genetics and Psychology.* Hillsdale, N.J.: Lawrence Erlbaum.

Plomin, R. and Daniels, D. (1987) Why are children in the same family so different from one another? *Behavioral Brain Sciences*, 10, 1–16.

Plomin, R., De Fries, J.C. and McClean, G.E. (1990) *Behavioral Genetics.* New York: W.H. Freeman.

Plotnick, R. (1992) Welfare and out-of-wedlock childbearing: Evidence from the 1980's. *Journal of Marriage and the Family*, 54, 23–35.

Post, F. (1994) Creativity and psychopathology. *British Journal of Psychiatry*, 165, 22–34.

Pound, J.F. (1972) An Elizabethan census of the poor. *University of Birmingham Historical Journal*, 7, 142–160.

Preston, S.H. (1974) Differential fertility, unwanted fertility and racial trends in occupational achievement. *American Sociological Review*, 39, 492–506.

Preston, S.H. and Campbell, C. (1993) Differential fertility and the distribution of traits: The case of IQ. *American Journal of Sociology*, 98, 997–1019.

Prins, M. (1994) Incidence and risk factors for acquisition of sexually transmitted diseases in heterosexuals with multiple partners. *Sexually Transmitted Diseases*, 21, 258–267.

Reed, E.W. and Reed, S.C. (1965) *Mental Retardation.* Philadelphia, Pa.: W.B. Saunders.

Retherford, R.D. and Sewell, W.H. (1988) Intelligence and family size reconsidered. *Social Biology*, 35, 1–40.

Retherford, R.D. and Sewell, W.H. (1991) Birth order and intelligence: Further tests of the confluence model. *American Sociological Review*, 56, 141–158.

Reynolds, C.R., Chastain, R.L., Kaufman, A.S. and McLean, J.E. (1987) Demographic characteristics and IQ among adults. *Journal of School Psychology*, 25, 323–342.

Rice, R.J., Roberts, P.L., Handsfield, H.H. and Holmes, K.K. (1991) Socio-demographic distribution of gonorrhea incidence. *American Journal of Public Health*, 81, 1252–1258.

Rindfuss, R.R., Bumpass, L. and John, C.S. (1980) Education and fertility: Implications for the roles women occupy. *American Sociological Review*, 45, 431–447.

Roberts, J.A.F., Norman, R.M. and Griffiths, R. (1938) Studies on a child population: Intelligence and family size. *Annals of Eugenics*, 8, 178–215.

Robins, L.N. and Regier, D.A. (1991) *Psychiatric Disorders in America*. New York: Free Press.

Rogers, M.F. (1987) Transmission of HIV infection in the United States. In B.K. Silverman and A. Waddell (eds.), *Report of the Surgeon General's Workshop on Children with HIV Infection and Their Families*. Washington, D.C.: Department of Health and Human Sciences.

Rokeach, M. (1973) *The Nature of Human Values*. New York: Free Press.

Roll, J. (1992) *Lone Parents in the European Community*. London: European Family and Social Policy Unit.

Rose, S., Kamin, L.J. and Lewontin, R.C. (1984) *Not in Our Genes*. London: Pantheon.

Rothstein, M.G., Paunonen, S.V., Rush, J.C. and King, G.A. (1994) Personality and cognitive ability predictors of performance in graduate business school. *Journal of Applied Psychology*, 86, 516–530.

Rowe, D.C. (1986) Genetic and environmental components of anti-social behavior: A study of 265 twin pairs. *Criminology*, 24, 513–532.

Rowe, D.C. (1994) *The Limits of Family Experience*. New York: Guilford Press.

Rowland, D.T. (1989) Who's producing the next generation? The parentage of Australian children. *Journal of the Australian Population Association*, 6, 1–17.

Rowntree, B.S. (1901) *Poverty: A Study of Town Life*. London: MacMillan.

Rubinstein, E. (1994) 200 years and out. *National Review*, 4 April, 20.

Runciman, W.G. (1964) Embourgeoisement. *Sociological Review*, 12, 158–176.

Rushton, J.P. (1994) *Race, Evolution and Behavior: A Life History Perspective*. New York: Transaction.

Rushton, J.P. and Bogaert, A.F. (1988) Race versus social class differences in sexual behavior: A follow-up of the r/k dimension. *Journal of Research in Personality*, 22, 259–272.

Rushton, J.P., Fulker, D.W., Neale, M.C., Nias, D.K. and Eysenck, H.J. (1986) Altruism and aggression: The heritability of individual differences. *Journal of Personality and Social Psychology*, 50, 1192–1198.

Rutter, M., Tizard, J. and Whitmore, K. (1981) *Education, Health and Behavior*. New York: R.E. Kreiber.

Saleeby, C. (1914) *The Progress of Eugenics*. London: Methuen.

Sano, T. (1974) Differences over time in intellectual ability. *Japanese Journal of Educational Psychology*, 22, 110–114.

Sarbin, T.R. (1970) The culture of poverty, social identity and cognitive outcomes. In V.L. Allen (ed.), *Psychological Factors in Poverty*. Chicago: Markham.

Scarr, S. (1969) Effects of birth weight on later intelligence. *Social Biology*, 16, 249–256.

Scarr, S. (1984) *Race, Social Class and Individual Differences in IQ.* London: Lawrence Erlbaum.

Scarr, S. and Weinberg, R.A. (1977) Intellectual similarities within families of both adopted and biological children. *Intelligence*, 1, 170–191.

Scarr, S. and Weinberg, R.A. (1978) The influence of family background on intellectual attainment. *American Sociological Review*, 43, 674–692.

Schofield, M. (1965) *The Sexual Behavior of Young People.* London: Longman.

Schulsinger, F. (1972) Psychopathy, heredity and environment. *International Journal of Mental Health*, 1, 190–206.

Schultz, F.R. (1983) Phenylketonuria and other metabolic diseases. In J.A. Blackman (ed.), *Medical Aspects of Developmental Disabilities in Children Birth to Three.* Iowa City: University of Iowa Press.

Schultz, H. (1991) Social differences in mortality in the eighteenth century. *International Review of Social History*, 36, 232–248.

Scottish Council for Research in Education. (1949) *The Trend of Scottish Intelligence.* London: University of London Press.

Shaikh, K. and Becker, S. (1985) Socio-economic status and fertility in rural Bangladesh. *Journal of Biosocial Science*, 17, 81–89.

Shaw, N. and Paleo, L. (1986) Women and AIDS. In L. McKusick (ed.), *What to Do about AIDS.* Berkeley, Calif.: UC Regents.

Shockley, W. (1974) Sterilisation—a thinking exercise. In C.J. Bajema (ed.), *Eugenics Then and Now.* Strondsburg, Pa.: Dowden, Hutchinson and Ross.

Shockley, W.B. (1972) Dysgenics, Geneticity and Raceology. *Phi Delta Kappan* (Special Supplement), 297–312.

Singapore (1980) *Census.* Singapore: Government Printing Office.

Sirkin, M. (1929) The relation between intelligence, age and home environment of elementary school pupils. *School and Society*, 30, 304–308.

Skipp, V. (1978) *Crisis and Development: An Ecological Case Study of the Forest of Arden 1570–1674.* Cambridge: Cambridge University Press.

Skodak, M. and Skeels, H.M. (1949) A final follow-up study of one hundred adopted children. *Journal of Genetic Psychology*, 75, 85–125.

Smith, J.D. (1994) Reflections on mental retardation and eugenics, old and new. *Mental Retardation*, 32, 234–238.

Smith, S. (1942) Language and nonverbal test performance of racial groups in Honolulu before and after a 14 year interval. *Journal of General Psychology*, 26, 51–93.

Soloway, R.A. (1990) *Demography and Degeneration.* Chapel Hill: University of North Carolina Press.

Sonko, S. (1994) Fertility and culture in sub-Saharan Africa: A review. *International Social Science Journal*, 46, 401–411.

Sorokin, P. (1927) *Social Mobility.* New York: Harper.

Spencer, H. (1874) *Study of Sociology.* London: MacMillan.

Statistics Netherlands. (1993) *Netherlands Fertility and Family Survey.* The Hague: Government Statistics Bureau.

Stevenson, T.H.C. (1920) Fertility of various classes in England and Wales from the middle of the 19th century to 1911. *Journal of the Royal Statistical Society*, 83, 401–442.

Stewart, L. and Pascual-Leone, J. (1992) Mental capacity constraints and the development of moral reasoning. *Journal of Experimental Child Psychology*, 54, 251–287.

St. Louis, M.E., Conway, G.A., Hayman, C.R., Miller, C., Petersen, L.R. and Dondero, T.J. (1991) Human immunodeficiency virus infection in disadvantaged adolescents. *Journal of the American Medical Association*, 266, 2387–2391.

Stone, L. and Stone, J.C.F. (1986) *An Open Elite?* Oxford: Oxford University Press.

Storfer, M.D. (1990) *Intelligence and Giftedness.* San Francisco: Jossey-Bass.

Sundin, J. (1992) Sinful sex: Legal prosecution of extra-marital sex in pre-industrial Sweden. *Social Science History*, 16, 99–128.

Sutherland, H.E.G. (1930) The relationship between intelligence quotient and size of family in the case of fatherless children. *Journal of Genetic Psychology*, 38, 161–170.

Sutherland, H.E.G. and Thomson, G.H. (1926) The correlation between intelligence and size of family. *British Journal of Psychology*, 17, 81–92.

Swan, A.V., Murray, M. and Jarrett, L. (1991) *Smoking Behavior from Preadolescence to Young Adulthood.* Aldershot: Avebury Press.

Swan, G.E., Cardon, L.R. and Carmelli, D. (1994) *The Consumption of Tobacco, Alcohol and Caffeine in Male Twins: A Multivariate Genetic Analysis.* Paper presented at the Annual Meeting of the Society for Behavioral Medicine, Boston.

Taeuber, I.B. (1960) Japan's demographic transition re-examined. *Population Studies*, 14, 28–39.

Tambs, K., Sundet, J.M., Magnus, P. and Berg, K. (1989) Genetic and environmental contributions to the covariance between occupational status, educational attainment and IQ: A study of twins. *Behavior Genetics*, 19, 209–221.

Tapper, A. (1990) *The Family in the Welfare State.* Perth: Australian Institute for Public Policy.

Taubman, P. (1976) The determinants of earnings: Genetics, family and other environments; a study of male twins. *American Economic Review*, 66, 858–870.

Teasdale, T.W. (1979) Social class correlations among adoptees and their biological and adoptive parents. *Behavior Genetics*, 9, 103–114.

Teasdale, T.W. and Owen, D.R. (1981) Social class correlations among separately adopted siblings and unrelated individuals adopted together. *Behavior Genetics*, 11, 577–588.

Teasdale, T.W. and Owen, D.R. (1984a) Social class and mobility in male adoptees and non-adoptees. *Journal of Biosocial Science, 16, 521–530.*

Teasdale, T.W. and Owen, D.R. (1984b) Heredity and familial environment in intelligence and education level—a sibling study. *Nature*, 309, 620–622.

Teasdale, T.W. and Owen, D.R. (1989) Continuing secular increases in intelligence and a stable prevalence of high intelligence levels. *Intelligence*, 13, 255–262.

Teasdale, T.W. and Sorensen, T.I. (1983) Educational attainment and social class in adoptees: Genetic and environmental contributions. *Journal of Biosocial Science*, 15, 509–518.

Teasdale, T.W., Sorensen, T.I. and Owen, D.R. (1984) Social class in adopted and non-adopted siblings. *Behavior Genetics*, 14, 587–593.

Tellegen, A. (1985) Structure of mood and personality and their relevance to assessing anxiety. In A.H. Tuma and J.D. Maser (eds.), *Anxiety and the Anxiety Disorders.* Hillsdale, N.J.: Lawrence Erlbaum.

Tellegen, A., Lykken, D.T., Bouchard, T.J., Wilcox, K.J., Segal, N.L. and Rich, S. (1988) Personality similarity in twins reared apart and together. *Journal of Personality and Social Psychology*, 54, 1031–1039.

Terman, L.M. (1922) Were we born that way? *World's Work*, 44, 660–682.
Terman, L.M. (1959) *The Gifted Group at Mid-life.* Stanford: Stanford University Press.
Terman, L.M. and Merrill, M.A. (1937) *Measuring Intelligence.* London: Harrap.
Thomson, G.H. (1946) The trend of national intelligence. *Eugenics Review*, 38, 9–18.
Thomson, G.H. (1949) Intelligence and fertility: The Scottish 1947 survey. *Eugenics Review*, 41, 163–170.
Thorndike, E.L. (1913) Eugenics. *Scientific Monthly*, 83, 128–140.
Thorndike, R.L. (1975) Mr. Binet's test 70 years later. *Educational Researcher*, 4, 3–7.
Thorne, J.O. and Collocott, T.C. (1984) *Chambers Biographical Dictionary.* Edinburgh: Chambers.
Thurstone, L.L. and Jenkins, R.L. (1931) *Order of Birth, Parental Age and Intelligence.* Chicago: University of Chicago Press.
Todd, G.F. and Mason, J.I. (1959) Concordance of smoking habits in monozygotic and dizygotic twins. *Heredity*, 13, 417–444.
Tomlinson-Keasey and Little, T.D. (1990) Predicting educational attainment, occupational achievement, intellectual skill and personal adjustment among gifted men and women. *Journal of Educational Psychology*, 82, 442–455.
Tuddenham, R.D. (1948) Soldier intelligence in World Wars I and II. *American Psychologist*, 3, 54–56.
Tupes, E.C. and Christal, R.E. (1961) Recurrent personality factors based on trait ratings. *USAF ASD Tech. Rep.* No. 57–125.
Tygart, C.E. (1991) Juvenile delinquency and number of children in a family. *Youth and Society*, 22, 525–536.
Udry, J.R. (1978) Differential fertility by intelligence: The role of family planning. *Social Biology*, 25, 10–14.
Ushijima, Y. (1961) Changes in IQ level. *Jidro Shinri*, 15, 629–635.
Vallot, F. (1973) Résultats globaux: Niveau intellectuel selon le milieu social et scolaire. In *Enquête Nationale sur le Niveau Intellectuel des Enfants d'age Scolaire.* Cahier No. 64. Paris: University of Paris Press.
Van Court, M. (1985) *Intelligence and Fertility in the United States.* M.Sc. Thesis, University of Texas.
Van Court, M. and Bean, F.D. (1985) Intelligence and fertility in the United States: 1912–1982. *Intelligence*, 9, 23–32.
Van Dusen, K.T., Mednick, S.A., Gabrielli, W.E. and Hutings, B. (1983) Social class and crime in an adoption cohort. *Journal of Criminal Law and Criminology*, 74, 249–269.
Vernon, P.E. (1951) Recent investigations of intelligence and its measurement. *Eugenics Review*, 43, 125–137.
Vernon, P.E. (1979) *Intelligence.* San Francisco: W.H. Freeman.
Vining, D.R. (1982) On the possibility of the re-emergence of a dysgenic trend with respect to intelligence in American fertility differentials. *Intelligence*, 6, 241–264.
Vining, D.R. (1995) On the possibility of the re-emergence of a dysgenic trend: An update. *Personality and Individual Differences*, 19, 259–265.
Vining, D.R., Bygren, L., Hattori, K., Nystrom, S. and Tamura, S. (1988) IQ/fertility relationships in Japan and Sweden. *Personality and Individual Differences*, 9, 931–932.
Vogler, G.P. and Fulker, D.W. (1983) Familial resemblance for educational attainment. *Behavior Genetics*, 13, 341–354.

Wallace, A.R. (1890) Human selection. *Popular Science Monthly*, 38, 90–102.

Waller, J. (1971) Differential reproduction: Its relation to IQ test scores, education and occupation. *Social Biology*, 18, 122–136.

Watanabe, Y. (1990) Demographic life stages of Japanese women: Cohort trends and socio-economic variations (in Japanese). *Journal of Population Problems*, No. 195, 49–58.

Webb, S. (1896) *The Difficulties of Individualism.* London: Fabian Society.

Weber, M. (1904) *The Protestant Ethic and the Spirit of Capitalism.* New York: Scribner.

Webster, R.A., Hunter, M. and Keats, J.A. (1994) Peer and parental influences on adolescents' substance abuse. *International Journal of the Addictions*, 29, 647–657.

Weinberg, R.A., Scarr, S. and Waldman, I.D. (1992) The Minnesota transracial adoption study: A follow-up of IQ test performance at adolescence. *Intelligence*, 16, 117–135.

Weiss, V. (1990) Social and demographic origins of the European proletariat. *Mankind Quarterly*, 31, 126–152.

Weissman, M.M., Myers, J.K. and Harding, P.S. (1978) Psychiatric disorder in a U.S. urban community. *American Journal of Psychiatry*, 135, 459–462.

Wellings, K., Field, J., Johnson, A. and Wadsworth, J. (1994) *Sexual Behavior in Britain.* London: Penguin.

West, D.J. and Farrington, D.P. (1971) *The Delinquent Way of Life.* London: Heinemann.

White, K.R. (1982) The relation between socio-economic status and academic achievement. *Psychological Bulletin*, 91, 461–481.

Whitehead, R.G. and Paul, A.A. (1988) Comparative infant nutrition in man and other animals. In R.G. Whitehead and A.A. Paul (eds.), *Proceedings of the International Symposium on Comparative Nutrition.* London: Libbey.

Wiggins, N., Blackburn, M. and Hackman, M. (1969) Prediction of the first year graduate success in psychology peer ratings. *Journal of Educational Research*, 63, 81–85.

Wikstrom, P.O. (1987) *Patterns of Crime in a Birth Cohort.* Stockholm: University of Stockholm Department of Sociology.

Willerman, L., Loehlin, J.C. and Horn, J.M. (1992) An adoption and a cross-fostering study of the Minnesota Multiphasic Personality Inventory Psychopathic Deviate Scale. *Behavior Genetics*, 22, 515–529.

Williamson, D., Serdula, M., Kendrick, J. and Binkin, N. (1989) Comparing the prevalence of smoking in pregnant and non-pregnant women. *Journal of the American Medical Association*, 261, 70–74.

Wilson, C. (1982) *Marital Fertility in Pre-industrial England, 1550–1849.* Ph.D. Thesis, Cambridge University.

Wilson, J.Q. and Herrnstein, R.J. (1985) *Crime and Human Nature.* New York: Simon and Schuster.

Winkleby, M., Fortmann, S. and Barratt, D. (1990) Social class disparities in risk factors for disease: Eight year prevalence patterns by level of education. *Preventive Medicine*, 19, 142.

Wong, D.H. (1980) *Class Fertility Trends in Western Nations.* New York: Arno Press.

Wood, J.L., Johnson, P.L. and Campbell, K.L. (1985) Demographic and endocrinological aspects of low natural fertility in highland New Guinea. *Journal of Biosocial Science*, 17, 57–79.

Wrigley, E.A. and Schofield, R.S. (1981) *The Population of England, 1541–1871: A Reconstruction.* Cambridge: Harvard University Press.

Wynne-Edwards, V.C. (1962) *Animal Dispersion in Relation to Social Behavior.* Edinburgh: Oliver and Boyd.

Xizhe, P. (1991) *Demographic Transition in China.* Oxford: Clarendon Press.

Yates, F. and Mather, K. (1963) Ronald Aylmer Fisher. *Biographical Memoirs of the Royal Society*, 9, 91–130.

Young, M. (1958) *The Rise of the Meritocracy.* London: Thames and Hudson.

Young, M. and Gibson, J.B. (1965) Social mobility and fertility. In J.E. Meade and A.S. Parkes (eds.), *Biological Aspects of Social Problems.* Edinburgh: Oliver and Boyd.

Zajonc, R.B. and Marcus, G.B. (1975) Birth order and intellectual development. *Journal of Personality and Social Psychology*, 37, 1325–1341.

Zazzo, R. (1960) *Les jumeaux, le couple et la personne.* Paris: Presse Universities de France.

Zhihui Y. (1990) Study on the fertility of the female-aged population above 60. *Population Research*, 7, 30–37.

Zill, N., Moore, K., Ward, C. and Steif, T. (1991) *Welfare Mothers as Potential Employees.* Washington, D.C.: Child Trends, Inc.

Zimmerman, S.E., Martin, R. and Vlahov, D. (1991) Aids knowledge and risk perceptions among Pennsylvania prisoners. *Journal of Criminal Justice*, 19, 239–256.

Index

About the Author

RICHARD LYNN is Director of the Ulster Institute for Social Research, Coleraine, Northern Ireland. He graduated in Psychology at the University of Cambridge and has held positions as Professor of Psychology at the Economic and Social Research Institute, Dublin, and at the University of Ulster. Among his earlier books are *Personality and National Character* and *Educational Achievement in Japan.*

ISBN 0-275-94917-6
90000>
EAN
9 780275 949174
HARDCOVER BAR CODE